Analog Circuits and Signal Processing

The *Analog Circuits and Signal Processing* book series, formerly known as the *Kluwer International Series in Engineering and Computer Science*, is a high level academic and professional series publishing research on the design and applications of analog integrated circuits and signal processing circuits and systems. Typically per year we publish between 5-15 research monographs, professional books, handbooks, and edited volumes with worldwide distribution to engineers, researchers, educators, and libraries. The book series promotes and expedites the dissemination of new research results and tutorial views in the analog field. There is an exciting and large volume of research activity in the field worldwide. Researchers are striving to bridge the gap between classical analog work and recent advances in very large scale integration (VLSI) technologies with improved analog capabilities. Analog VLSI has been recognized as a major technology for future information processing. Analog work is showing signs of dramatic changes with emphasis on interdisciplinary research efforts combining device/circuit/technology issues. Consequently, new design concepts, strategies and design tools are being unveiled. Topics of interest include: Analog Interface Circuits and Systems; Data converters; Active-RC, switched-capacitor and continuous-time integrated filters; Mixed analog/digital VLSI; Simulation and modeling, mixed-mode simulation; Analog nonlinear and computational circuits and signal processing; Analog Artificial Neural Networks/Artificial Intelligence; Current-mode Signal Processing; Computer-Aided Design (CAD) tools; Analog Design in emerging technologies (Scalable CMOS, BiCMOS, GaAs, heterojunction and floating gate technologies, etc.); Analog Design for Test; Integrated sensors and actuators; Analog Design Automation/Knowledge-based Systems; Analog VLSI cell libraries; Analog product development; RF Front ends, Wireless communications and Microwave Circuits; Analog behavioral modeling, Analog HDL.

Jonah Van Assche • Georges Gielen

Event-Driven Circuit Architectures for Scalable and Adaptive Sensor Readout

Springer

Jonah Van Assche
Institute of Smart Sensors
University of Stuttgart
Stuttgart, Germany

Georges Gielen
ESAT-MICAS
KU Leuven
Leuven, Belgium

ISSN 1872-082X ISSN 2197-1854 (electronic)
Analog Circuits and Signal Processing
ISBN 978-3-031-97284-3 ISBN 978-3-031-97285-0 (eBook)
https://doi.org/10.1007/978-3-031-97285-0

This Springer imprint is published by the registered company Springer Nature Switzerland AG
The registered company address is: Gewerbestrasse 11, 6330 Cham, Switzerland

Preface

The continuous monitoring of vital body signs such as the electrocardiogram (ECG) and electroencephalogram (EEG) signals is crucial for the early detection of health issues. Achieving long-term monitoring with wireless connectivity, however, poses significant challenges. Sensing devices for vital signals must operate for extended periods of time on small, portable batteries, making an ultra-low energy consumption a critical concern. Often, on-chip data reduction through digital signal processing is necessary to meet the energy budget constraints. While digital signal processing can extend the battery life, innovative sensor architectures are needed for even larger power savings.

This book addresses the need for energy-efficient sensing devices by focusing on event-driven processing in the mixed-signal domain. By investigating event-driven sensing techniques that adaptively adjust the sampling rate based on the signal activity of time-sparse signals such as the ECG or action potentials (AP), this work aims to minimize the power consumption of the sensing device as well as the transmission power. This will be explored by investigating event-driven level-crossing ADCs (LCADCs).

First, the advantages of LCADCs are thoroughly investigated. The circuit limitations of LCADCs are analyzed by means of a behavioral MATLAB model. Next, an analytical model for the LCADC power consumption is derived. Two comparisons are performed between the LCADC architecture and a conventional Nyquist-rate ADC with digital compression: one at the ADC level and one at the system level. At the ADC level, LCADC power models are compared with power models of a SAR ADC, the most commonly used ADC architecture in conventional biomedical systems. The aim is to determine the conditions under which LCADCs are more power efficient. At the system level, LCADCs are compared with SAR ADCs with digital compression to assess the system-level benefits of the LCADC topology. LCADCs prove to be more power-efficient compared to SAR ADCs for low to medium resolutions (e.g., up to 6.8 bits for ECG and EEG in a 40 nm CMOS technology). Conventional LCADCs are limited by the fixed power consumption of the continuous-time (CT) comparator and, if applicable, the time-to-digital converter (TDC). At system level, LCADCs that directly stream the data off chip can significantly reduce the data transmission power, up to two orders of magnitude for a 5-bit resolution, and achieve an order of magnitude power reduction for an 8-bit resolution. However, for applications where the LCADC output needs to be retimed with a TDC, no significant advantage over digital compression is observed. Hence, the following parts of the book focus on improved LCADC architectures.

Next, a first IC design is presented that addresses the limitations of conventional LCADCs by introducing an LCADC with discrete-time (DT) comparators and adaptive clocking. By using DT comparators instead of CT comparators, a significant power consumption reduction can be achieved. To exploit the time-sparse properties of the signals, an adaptive-clocking strategy is employed, where the ADC sampling rate is adjusted using an on-chip clock generator and off-chip event logic on an FPGA. This adaptive-clocking strategy can enable a 10–100× power saving in the comparator. A prototype chip is fabricated in a 40 nm CMOS technology, with an 8-bit quantizer resolution and 15-kHz bandwidth. The fabricated LCADC exhibits a variable power consumption, ranging from 0.92 μW for an ECG input to 5.38 μW for a full-scale 15-kHz sine input. The design achieves an effective number of bits (ENOB) of 10.4 bits, a best-case Walden signal-to-noise-and-distortion ratio (SNDR)/dynamic range (DR) figure of merit (FoM) of 138/119.8 fJ/conversion step, and a best-case Schreier SNDR/DR FoM of 158.85/152 dB. The system-level benefits of this topology are demonstrated through a 30% data reduction for ECG applications.

To build upon the first prototype, a second IC is then presented, the FREYA SoC. This IC extends the synchronous LCADC to a multi-channel analog-to-spike converter (ASC), which generates spiking inputs for an on-chip spiking neural network (SNN) processor. The FREYA prototype chip consists of an 8-channel, time-division-multiplexed ASC and an integrated SNN processor. Thanks to the adaptive system clock and the multiplexing, the proposed ASC can reuse the same hardware for each channel. The ASC also features a programmable adaptive-resolution algorithm, allowing the tuning of the spike encoding to any signal of interest. A custom training framework for the SoC can efficiently and rapidly tune the ASC and SNN toward a specific application, reducing the required training epochs of the SNN by 69% compared to a full grid search. The framework has been demonstrated for the QRS-labeling task based on ECG, achieving a maximum accuracy of 98.67%. A prototype IC fabricated in a 40 nm CMOS technology demonstrates significant improvements in chip area efficiency, with a per-channel active area of only 0.023 mm^2/channel (0.184 mm^2 in total), a 7-fold improvement over similar works. For the QRS-labeling task, the system consumes 20.8 μW/channel and 13.29 μJ/classification, with an SNN clock running at 15 MHz (VDD at 0.56 V), and achieves a latency of 80 ms.

Tübingen, Germany
Leuven, Belgium
February 2025

Jonah Van Assche
Georges Gielen

Declarations

Competing Interests The authors have no competing interests to declare that are relevant to the content of this manuscript.

Contents

Acronyms

ADC	Analog-to-Digital Converter
AER	Address-Event Representation
AFE	Analog Front-End
AP	Action Potential
ASC	Analog-to-Spike Converter
ASIC	Application-Specific Integrated Circuit
AWG	Arbitrary Waveform Generator
BLE	Bluetooth Low Energy
CCIA	Capacitively coupled Chopper-stabilized Instrumentation Amplifier
CDAC	Capacitive DAC
CM	Common Mode
CMOS	Complementary Metal-Oxide-Semiconductor
CMRR	Common Mode Rejection Ratio
CNN	Convolutional Neural Network
CR	Compression Ratio
CS	Compressive Sensing
CT	Continuous Time or Continuous-Time
DAC	Digital-to-Analog Converter
DEM	Dynamic Element Matching
DNL	Differential Non-linearity
DSL	DC-Servo Loop
DSP	Digital Signal Processor
DR	Dynamic Range
DRL	Driven Right Leg
DT	Discrete Time or Discrete-Time
ECG	Electrocardiogram
EEG	Electroencephalogram
EMG	Electromyogram
ENOB	Effective Number Of Bits
EOG	Electrooculogram
EOV	Electrode Offset Voltage

FFT	Fast Fourier Transform
FoM	Figure of Merit
FPGA	Field Programmable Gate Array
HBC	Human Body Communication
IA	Instrumentation Amplifier
IC	Integrated Circuit
INL	Integral Non-linearity
IoT	Internet of Things
IRN	Input-Referred Noise
LCADC	Level-Crossing ADC
LCS	Level-Crossing Sampling
LFP	Local Field Potential
LNA	Low-Noise Amplifier
LSB	Least Significant Bit
ML	Machine Learning
MOMCAP	Metal-Oxide-Metal Capacitor
MOS	Metal-Oxide-Semiconductor
MOSFET	Metal-Oxide-Semiconductor Field-Effect Transistor
MSE	Mean Square Error
NBW	Equivalent Noise Bandwidth
NEF	Noise Efficiency Factor
NMOS	N-type Metal-Oxide-Semiconductor
NN	Neural Network
OSR	Oversampling Ratio
PDF	Probability Density Function
PEF	Power Efficiency Factor
PMOS	P-type Metal-Oxide-Semiconductor
PSD	Power Spectral Density
RMS	Root-Means-Square
SAR ADC	Successive-Approximation ADC
SBW	Signal Bandwidth
SER	Signal-to-Error Ratio
SNDR	Signal-to-Noise-and-Distortion Ratio
SNN	Spiking Neural Network
SNR	Signal-to-Noise Ratio
SoC	System-on-Chip
SQNR	Signal-to-Quantization-Noise Ratio
SRAM	Static Random Access Memory
SVM	Support Vector Machine
TDC	Time-to-Digital Converter
TF	Transfer Function
TX	Transmitter
Vpp	Peak-to-Peak Voltage
ZOH	Zero-Order-Hold

Chapter 1
Introduction

Sensors have become ubiquitous in the daily life of humans. They are used in various use cases, ranging from smart home assistants such as Alexa to (self-driving) cars to smartphones and fitness trackers, etc. In all these applications, information such as sound, vision, health status, etc. is transformed from an initial analog signal into digital data, which can be processed by algorithms, leading to insights and/or actions, much like a person uses his senses to navigate through the world.

This book focuses on the design of integrated sensor interfaces. These are the electronic circuits that convert the analog sensor signals into digital data. In this introductory chapter, the position of sensors in today's global IT and electronic infrastructure will be sketched, and some background regarding sensors and their electronic interfaces will be given. Then, several challenges that future sensor systems face will be identified, before defining the focus of this work. Finally, the book outline will be discussed, highlighting the main topics covered in the book and the original contributions of the work.

1.1 Smart Sensor Interfaces for the Edge

Over the past 75 years, progress in electronics has been fueled by Moore's law [87]. This economic observation, first made by Gordon Moore, predicted in 1965 that each 1.5 years, the amount of transistors that can be integrated on a silicon-chip doubles or that for the same amount of devices required for a functionality, the chip area (and hence the cost) can be scaled down [87]. This initial observation of Moore later became the driving force for the semiconductor industry and has led to a tremendous increase in the use of electronics in our society, which could be

J. Van Assche, G. Gielen, *Event-Driven Circuit Architectures for Scalable and Adaptive Sensor Readout*, Analog Circuits and Signal Processing,
https://doi.org/10.1007/978-3-031-97285-0_1

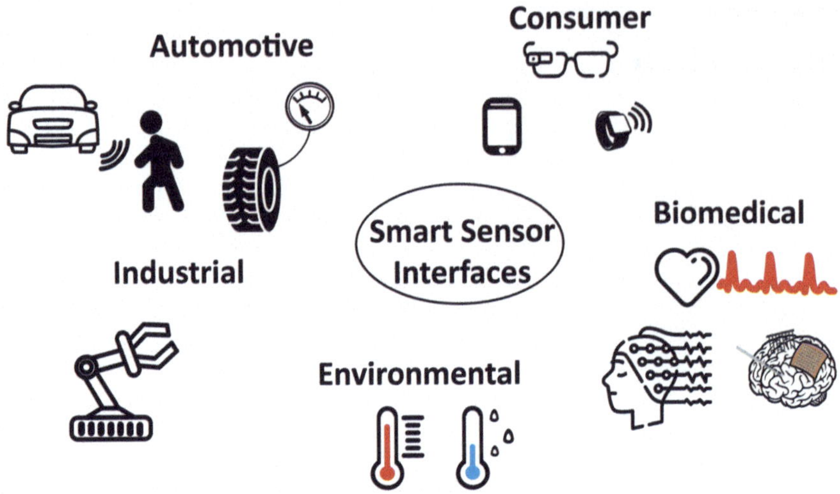

Fig. 1.1 Overview of different sensor application fields

made ever smaller and more efficient every 1.5 years.[1] Sensors, being electronic devices and providing inputs to electronic systems, naturally benefited from this trend. Today, almost any electronic system contains multiple sensors.

Figure 1.1 shows a couple of application fields that rely heavily on various sensors. They include:

- *Automotive*: Modern cars contain more than hundreds of sensors [3]. These can range from pressure sensors to monitor tire pressure, over temperature sensors that monitor the cabin temperature, to smart radar systems that help with, e.g., pedestrian detection, etc. With the increasing electrification of cars, the amount of sensors per car will only increase ever more [3].
- *Consumer*: Consumer electronics (smartphones, smartwatches, etc.) use various low-power sensors to improve the user interaction and the functionality of such devices. For example, the iPhone 15 contains multiple cameras, a barometer, a gyroscope, an accelerometer, etc. [2].
- *Biomedical*: To keep healthcare costs under control and to better diagnose patients, there is a push in the medical field to monitor patients not only in a controlled hospital environment (which is usually done with bulky, static instrumentation setups like scanners) but also in the home situation [5]. This requires wearable health devices, such as electrocardiogram (ECG) patches, electroencephalogram (EEG) headsets, etc. More futuristic devices, such as brain-computer interfaces, may (in the near future) also be used to monitor more complex physiological processes such as neural diseases like epilepsy [136].

[1] This rate later on slowed down to every 2 years.

- *Environmental*: For agriculture, aviation, urban development, etc. monitoring the environment is crucial to ensure the quality of crops, the safety of planes, the quality of the air, etc. To accurately measure environmental parameters that are of interest for such applications (temperature, pressure, humidity, etc.) in real time, environmental sensors play a key role [84].
- *Industrial*: Factories require that the production process runs smoothly, with ideally no interruption by defective machines, and for safety, a machine should be able to directly shut down in case of an emergency situation [3]. Sensors allow to monitor the condition of machines in real time, such that there can be an automatic intervention once a machine starts to operate incorrectly. Sensors can also automate many steps in the production process (e.g., quality control), which ultimately makes the production process more efficient.

Sensors thus enable a plethora of applications that require the knowledge about a certain (physical) quantity and that need to process the information embedded in that signal into insights [22, 83, 134]. An illustrative overview of some example sensor systems is given in Fig. 1.2.[2] A sensor system has four main components [83]. The first component is the physical quantity. This is the signal that the sensor system tries to capture. It can range from quantities such as heat or pressure to electrophysiological signals such as the ECG and EEG to photons (from, e.g., an LED or natural light). The second component of the system is the transducer, which converts the physical quantity to a quantity suited for electronics. This can be a voltage, a current, or a charge or by means of an active transducer also a resistance, capacitance, or inductance. Typically, this electronic quantity is a continuous, analog signal. Transducers are available in many forms and can be photodiodes, micro-electromechanical systems, electrodes, etc. The third component of the sensor, the sensor readout circuit, converts the analog input signal from the transducer into a digital representation. Typically, the sensor readout circuit consists of an amplifier to amplify the small input signal from the transducer, a filter to remove potential interference signals and an analog-to-digital converter (ADC) to convert the analog signal in digital bits. The last component of the sensor system is the signal processing, which converts the raw signal data into insights and/or. This signal processing can range from simple processing algorithms like thresholding to complex classification algorithms using techniques such as neural networks.

An important remark on the division made in the previous paragraph is that the collection of the raw data and the signal processing does not need to happen physically in the same electronic system. Depending on the application at hand, the signal processing can happen close to the sensor readout, or the raw sensor data can be collected in a central place (which is called a cloud server), where a powerful processing algorithm processes the data of many connected sensors [22, 134]. Such a network of connected sensors (which are referred to as sensor

[2] Note that this figure is far from a complete overview of all types of sensor systems and merely gives an illustrative view of how a sensor system functions. A complete review of all various sensor concepts is beyond the scope of this work.

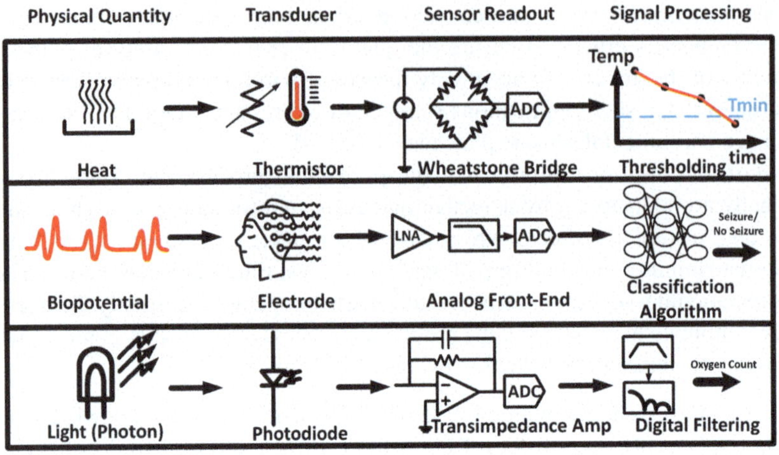

Fig. 1.2 Illustrative overview of some complete sensor chain, from physical quantity to signal processing. Given are the examples of a temperature sensor system, a biopotential (EEG) sensing system, and an optical system for photoplethysmogram measurements

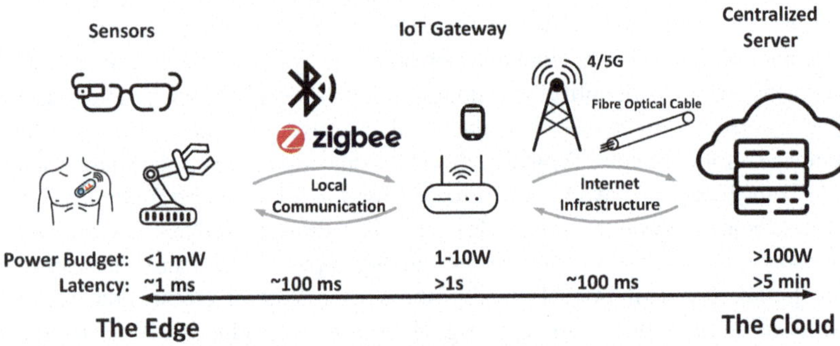

Fig. 1.3 The IoT ecosystem from devices on the edge (the various sensors and local base stations) to a centralized cloud server. Indicated are typical latency figures to transfer data from the edge toward the cloud servers, as well as the power budget available in each part of the network

nodes) is called the Internet of things (IoT). In Fig. 1.3, an overview of the IoT ecosystem is given. Sensor devices are said to be on the "edge" of the network, i.e., the point of access where data from the physical world are collected [22, 134]. The edge of the network consists often of many distributed sensors. The data from these sensors are then locally collected and transmitted onto the cloud. The device that takes care of this local data collection is called the IoT gateway or the base station and can, for example, be a smartphone or a laptop [22, 134]. Sensors communicate with the gateway by means of short-range communication protocols, such as Bluetooth Low Energy (BLE), ZigBee, human body communication (HBC), or WiFi [20, 41, 124, 187]. The gateway itself communicates with the cloud server by means of long-range communication technologies such as 4G/5G or an optical network. Indicated in Fig. 1.3 is also the power budget available in each part of the IoT network, as well as the latency for each data transfer. Generally speaking, the cloud servers have much more powerful processing capabilities, but it takes more time to gather, transmit, and process those data, increasing the latency. Reversely, processing in the edge needs to be low power, but it can happen at much lower latency [22, 124, 134]. In the next section, several limitations of current IoT nodes will be discussed, which will highlight why computing in the edge is increasingly becoming of interest.

1.2 Sensor System Challenges

The previous section has highlighted how the use of sensors can improve many aspects of our daily lives. By using connected networks of IoT nodes, a lot of information can be gathered, which can be used in various decision-making processes. The number of IoT devices that is being used is still increasing. In 2025, an expected 25 billion devices will be in use worldwide (see Fig. 1.4) [142]. With this expected increase in IoT devices, there also come many challenges, which need to be addressed to ensure sustainable growth of this market. To make these challenges more tangible, in this section four illustrative examples of sensor systems (depicted in Fig. 1.5) will be used to highlight these challenges. The use cases considered are a wearable ECG patch with a BLE transmitter, a multichannel (eight channels) EEG recording array with BLE transmitter, a multichannel implantable neural recording and stimulation array (256 readout channels) with HBC link, and a wireless security camera that streams images over a WiFi link.

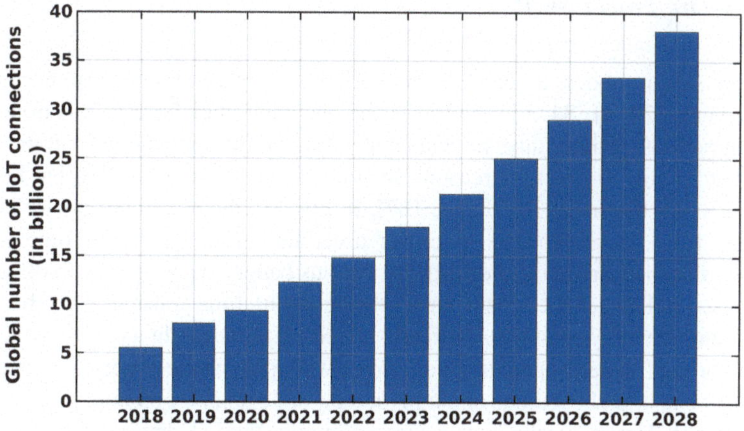

Fig. 1.4 Predicted amount of IoT connections (= amount of devices connected to an IoT node) in billions; data taken from [142]

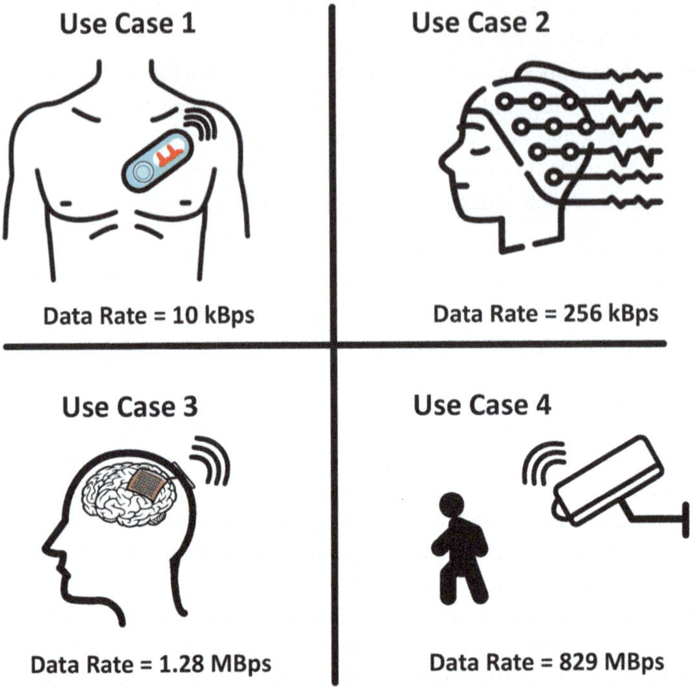

Fig. 1.5 Illustrative use cases of wireless sensor systems: a wearable ECG patch for health monitoring, a multichannel (eight channels) EEG recording array, a multichannel implantable neural recording and stimulation array (256 readout channels), and a wireless security camera

1.2.1 The Power Wall

Wireless sensor nodes in an IoT network have stringent power consumption requirements [20, 41, 84, 124, 187]. On the one hand, they have to function often for long periods of time (days to years depending on the application). On the other hand, due to cost or practical reasons such as physical size, sensors often have to work on small batteries [20, 78, 84, 187]. In Fig. 1.6, the maximum power budget available for a sensor node is given for a certain sensor battery lifetime (if the battery's self-discharge is neglected), for various battery types. The type of battery that is used influences the sensor node's operating lifetime considerably. However, the amount of energy available is directly traded off with weight and size, which are also important from an application point of view. Replacing the battery or recharging the sensor can often be tedious or simply not possible (e.g., for an implant or remote sensor node). Therefore, the power consumption of the IoT devices must meet the stringent power constraints set by the battery type and the required lifetime [41]. Another trend that can be noted, and that contrasts with the low-power requirements, is that sensor nodes more and more have to support an increasing number of readout

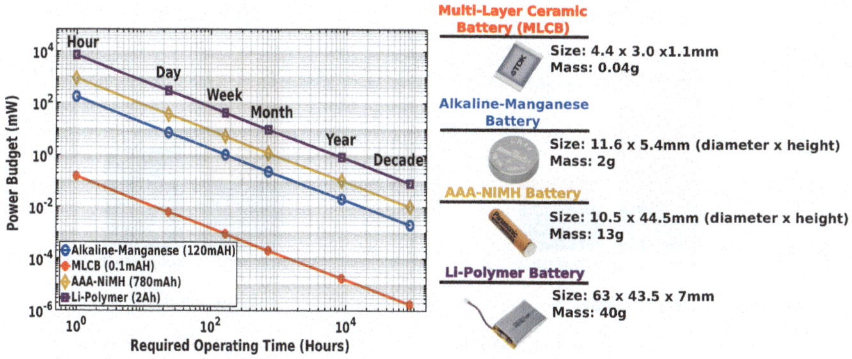

Fig. 1.6 Available power budget of a sensor node, as a function of the required operating time of the sensor node, for different battery types

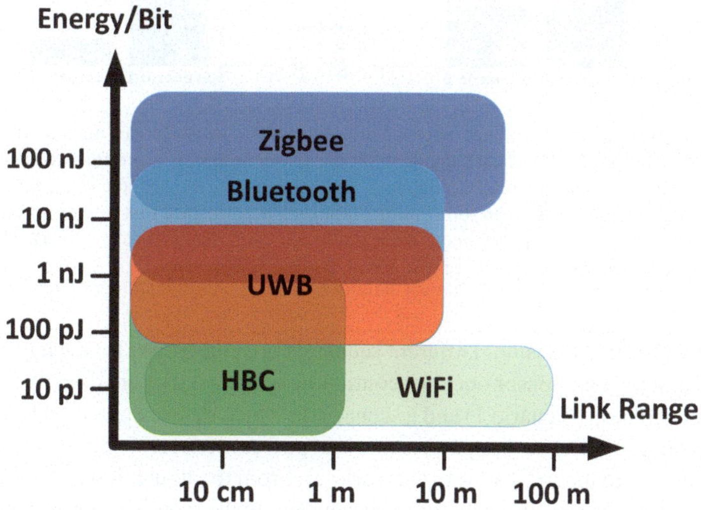

Fig. 1.7 Energy efficiency and typical link range of various short-range communication protocols

channels [136, 183]. Increasing the amount of sensor channels generates richer sensor data, which can be used for better decision-making, but increases the power of the sensor node if no additional circuit and system innovations are added to the sensor devices [183]. Hence, to continue both these trends, more intelligent and more sophisticated sensor design is required.

Most power of the sensor devices in the edge is consumed in the wireless transmission to the IoT gateway [21, 41, 48, 124, 155, 183]. Figure 1.7 shows the typical energy efficiency for various short-range communication protocols that are used to connect the sensors to a base station. For the four use cases of Figs. 1.5, 1.8 plots the power consumption of the sensor system and the power consumed

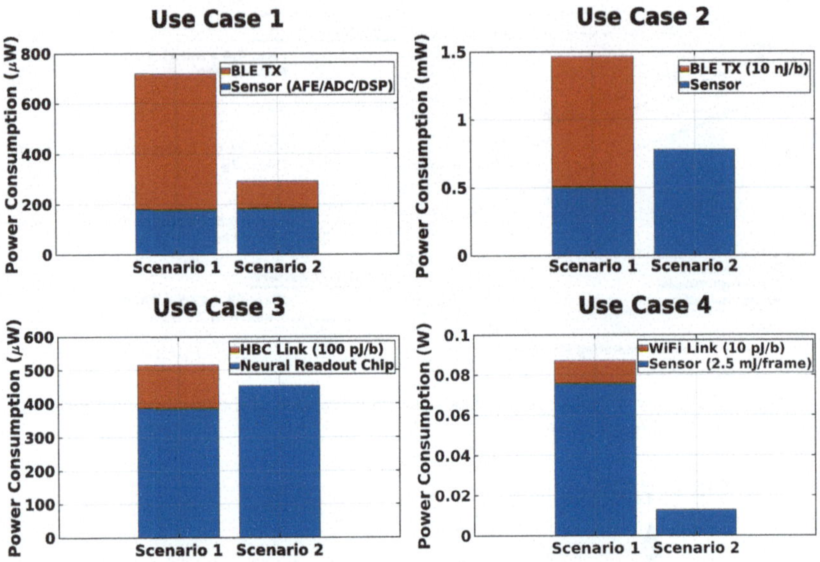

Fig. 1.8 Overview of the estimated power consumption of a sensor node, for the different use cases discussed in Fig. 1.5. For each application, the power budget is given for two scenarios. In the first scenario, the sensor works continuously and streams all raw data to a cloud server. In the second scenario, some on-chip (pre)processing of the data is performed first (use cases 1–3), or the sensor node works in a low-power mode until a relevant event happens (use case 4). The power numbers used are taken from literature [97, 136, 179, 182, 187]

by the wireless transmission. Different scenarios are considered for each use case: a scenario in which the sensor operates continuously and all data are directly streamed off the sensor node (scenario 1) and a scenario in which data are processed locally on the sensor (and/or the sensor operates in a low-power mode) and only processed data are transmitted to the IoT gateway (scenario 2). From the figure, it is clear that when the sensor systems continuously transmit raw data to the cloud, a significant amount of power is consumed in data transmission. In contrast, on-chip preprocessing of the data reduces significantly the power of the transmitter, due to the reduced data rate that the sensors need to transmit. For use case 4, it is also clear that operating in a low-power mode when no events of interest take place can help to reduce the power of the entire sensor system (both the sensor and the transmitter) significantly.

1.2.2 The Latency Wall

Sensor devices provide information about physical quantities, which can be used to make decisions or actions. The time between capturing the sensor data and such action is called the latency of the system; it is the minimum time needed to act on incoming changes in the sensor signal [22, 134]. Figure 1.3 highlighted for each component in the IoT network how much latency it introduces. Clearly, the further the data has to travel from the edge of the network, and thus the larger the gap between the source of the data and the place where the processing takes place, the larger the latency. Depending on the application at hand, latency can be a critical parameter. For example, in use case 1, if a patient who is monitored by the ECG patch has a sudden cardiac arrest, an ultrafast response is required. The latency that is introduced by sending the data to the cloud server might be fatal to the patient. Similarly, for use cases 2 and 3, if a patient is monitored for, e.g., epilepsy, if a seizure onset is detected, a neurostimulator needs to quickly suppress this upcoming seizure; otherwise the patient goes into a seizure state [124, 136]. Doing off-sensor processing of the neural signals can yield a 5–20 seconds longer latency [124], which is not tolerable for such applications. As a last example, for applications like use case 4, in which the processing of the camera images is directly performed on the sensor nodes, a significant reduction in latency can be noted (e.g., up to 5× according to [134]). A low latency can be required for such applications since security cameras may require a fast detection of intruders. Other imaging applications include, for instance, the face ID login on a smartphone. For the comfort of the user, the face ID recognition should work instantly, without having to wait for the cloud server to unlock the smartphone.

1.2.3 The Memory/Storage Wall

The previous section on the Power Wall has discussed how transmitting the raw data directly from the sensor node to the cloud puts a burden on the power consumption requirements of these devices. On the cloud side of the IoT network, there are also various challenges with the ever increasing amount of sensor data. With more and more connected devices (see Fig. 1.4), the requirements on the server side of the network also need to increase, as servers need to process and store increasingly more data coming from the IoT nodes. It is estimated that by 2025 all IoT devices will generate an aggregate amount of 79.4 zettabytes per year [53]. This increase in server storage requirements entails a large hardware and energy cost, which both lead to high economic costs for cloud service providers. Moreover, there is also an environmental price to pay. In [85], it was estimated that in 2025, the total amount of energy that will be consumed by data centers in the European Union will be over 90 TWh each year. The same study also found that to support this increase in data centers, more than 88 million SSD units would be required to facilitate all these data centers (compared to over only 1.4 million in 2010) [85]. This expected increase in the amount of data centers and data traffic will produce up to 3.5 % of the global carbon emissions by 2025 [85].

Another issue on the server side is that the data that are collected are not being used properly: Many sensor data that are collected are not used for any actual processing, or the data that are gathered are not used beyond simple tasks such as monitoring a certain process [78]. To fully enable the value of the IoT, this waste in data should be addressed. Moreover, collecting many data that need to be stored and not using those data leads to a waste of resources and consumes excessive energy. With the large environmental cost that is related to this energy consumption, the need for efficient storage/management of IoT data is clear. For example, in use cases 1, 2, and 3, collecting days long of ECG or neural data without properly processing these data and extracting information from it are not useful, because it is difficult for a doctor to check days worth of patient records. In contrast, when the sensor can already annotate parts of the signal, i.e., indicate which part of the signal might be of interest to the doctor, these data become a lot more useful. Similarly, in use case 4, while recordings can be stored temporarily, manually checking these data can be a tiresome work, and often video recordings are simply deleted after a certain time. As an alternative, some form of processing can determine which segments of a recording are interesting for the user, e.g., which part of the recording does not contain redundant data, such that only these data can be stored. Given that processed data are more useful and lead to relaxed server side requirements and since the (initial) processing can also be done on the sensor devices, the question can thus be asked if collecting and transmitting all these raw data makes sense in the first place.

1.2.4 The Privacy/Security Wall

A last limitation of current IoT nodes is the limited security and privacy that such devices offer. As discussed in the previous section, IoT nodes have a limited power budget and generate an abundance of data that are transmitted to a cloud server. These data can reveal much personal information on, for instance, the medical condition of users, critical infrastructure data, etc. [22, 27, 78, 134]. The more such data have to be transmitted from one device to another, the more points exist in which these data can be intercepted by a potential malicious third party [78]. Figure 1.9 gives an overview of how an IoT node can be breached by an external attacker. Each point of the interconnected node can be targeted. At the edge of the network, the attacker can try to intercept the sensor data (by means of, e.g., an insecure connection) and steal valuable information [22, 27, 78, 134], or the attacker can try and disable the sensor node by a physical attack [27], aiming, for instance, to sabotage or destroy the network infrastructure. The IoT gateways can also be subject to such security breaches, or a third party can try and pretend to be one of the sensor nodes and send spoof data to the cloud [27]. This could be done, for example, to mask the fact that a sensor node has been attacked or to disrupt the operation of the network by feeding false data to the cloud [27, 78]. A last place to attack is the server that stores the data from the IoT nodes. While typically these servers can be protected by a powerful firewall, if an attack on the server is successful, a massive data leak may occur, or the IoT network could go offline if the server is shut down.

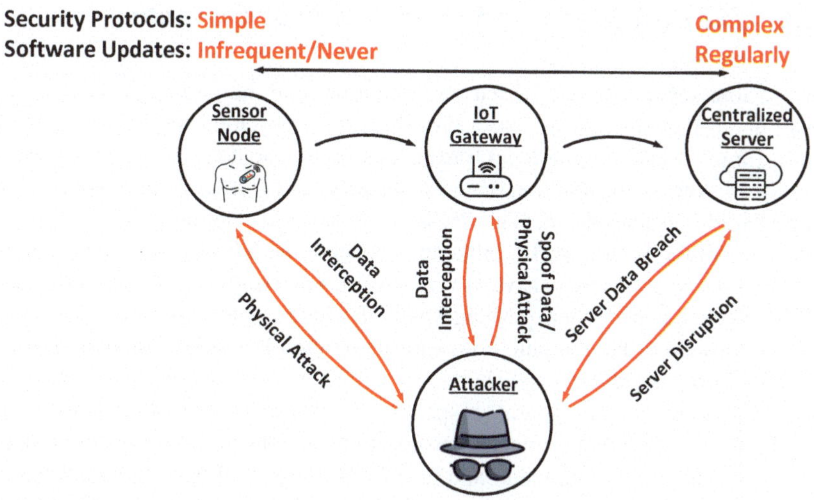

Fig. 1.9 Points of possible security breaches throughout the edge-to-cloud information stream

These examples highlight why there is a need to invest in a secure IoT network. On the sensor side of the network, some precautions can be taken. First off, data should be encrypted before transmission to the cloud. Encrypting data, however, comes at a cost, as the encryption algorithm causes a power overhead [27, 134, 141]. Hence, lightweight security protocols are needed to protect the sensor node [27, 141]. Moreover, once a sensor node is deployed, it is difficult to update the sensor's encryption protocol [27], as this may have been hard-coded on the sensor chip [141] or it is impractical to update the software of so many devices in the field [27]. Remarkably, not much research that focuses on circuit design for sensors prioritizes security of the sensor as a requirement. Hence, many new sensor readout circuits do not include any form of hardware security. Besides encrypting the data from the sensor node, another way to limit the risk of interception is to reduce the amount of points in which the data can leak [27, 78, 134]. For example, personal data regarding the health of a patient should only be transmitted to a server if something significant (e.g., a cardiac arrest, seizure episode, abnormal heart rhythm, etc.) is happening to the patient (for use cases 1–3); there is no need to transmit data if the patient is healthy.[3] Some information might also only be used locally, without the need to send it further to the cloud for storage [27, 134]. For use case 4, video images could be encrypted and only sent in case something significant (e.g. an intruder entering a building) is occurring, to keep a third party from tapping into the video stream.

[3] However, due to reliability reasons, probably sometimes data need to be transmitted even when no significant events are occurring, to check if the sensor is still functioning properly.

1.2.5 The Case for Smart Sensors and Edge Computing

The previous sections have all discussed limitations and shortcomings of current IoT sensor nodes that stream raw sensor data directly to the cloud and have highlighted that in some scenarios more local (pre)processing of the data can be advantageous in terms of power consumption, latency, security, and storage requirements. Computing locally on sensor data is also referred to as edge computing, i.e., computing in the edge of the IoT network. Unlike the conventional IoT nodes, edge-computing devices show lower latency, have lower transmission bandwidth requirements, and can be more secure. When pushing (part of) the computing to the edge of the network, however, the compute power available (see Fig. 1.3) is limited. There is thus an enormous (research) challenge to achieve similar sensing performance as traditional cloud computing while respecting the power budgets of the IoT devices. To achieve this goal, capturing and processing the data in the sensor node needs to be made much more efficient, and novel paradigms will need to be adapted. A conventional sensor nowadays captures data at a fixed rate, agnostic of the signal. Many sensors, however, have data that are sparse in nature, e.g., that only have short periods of actual information content. To enable more efficient data capturing and processing, exploiting this sparse nature can lead to large system power savings.

1.3 Book Outline

The focus of this book is the design of smart and low-power integrated sensor front-ends for future edge-computing devices, in particular biomedical sensor systems for biosignals such as ECG, EEG, etc. This work covers both the design of intelligent sampling front-ends and processing circuits for on-chip signal processing. The goal is to ***develop prototype readout chips that can achieve a lower sensor system power consumption and latency compared to current state-of-the-art implementations***, to enable future edge-computing devices. To reach this goal, event-driven sampling and processing techniques that exploit the sporadic activity of the biomedical sensor signals will be investigated. The structure of the book is as follows:

- In Chap. 2, a literature survey will be performed about biomedical sensor readout systems. After reviewing the major circuit building blocks for such systems, the chapter will discuss various techniques to lower the system data rate. Both classical digital compression and digital feature extraction and classification will be discussed. Next, techniques that compress the data in the analog and mixed-signal domain, such as compressive sensing and event-based sampling, will be reviewed. Level-crossing sampling in combination with event-driven processing emerges as a promising technique to achieve low system power consumption. Therefore, the following chapters will focus on this type of sensor readout.
- Chapter 3 will evaluate the performance of event-driven level-crossing sampling at the ADC level and the system level. First, system-level behavioral simulations

for level-crossing ADCs (LCADCs) will be done to assess the benefits and limitations of this topology. Second, the derivation of a power consumption model for this architecture will follow. Third, the LCADC will be compared to the classical SAR ADC, both at the ADC level and at the system level. From this comparison, the major bottleneck of the LCADC will become clear: the continuous-time comparator. An operating region will be identified in which the LCADC shows advantages over the SAR ADC.

- Chapter 4 will use the insights from the behavioral models and the analyses from Chap. 3 to propose a new LCADC design. First, the chapter will introduce a new LCADC topology which improves upon the classical LCADC topology. Second, a detailed circuit implementation will be described. Third, the measurement results from a prototype IC in 40 nm CMOS will be presented. The design achieves 10.4 maximum effective number of bits, for a 15-kHz bandwidth, with scalable power consumption (0.92 μW for an ECG signal to 5.38 μW for a full-scale 15-kHz sine wave).
- Chapter 5 will build further on Chap. 4. To push further the benefits of the LCADC to the system level, an event-driven system-on-chip (SoC) will be proposed, with an 8-channel level-crossing sampling analog-to-spike converter (ASC) and an integrated spiking neural network (SNN) processor. First, the LCADC concept of Chap. 4 will be extended to an area-efficient multichannel ASC implementation. The detailed circuit implementation will be presented, and a new dynamic element matching technique will be introduced. Second, a custom offline training framework to operate the SoC will be introduced, to optimally configure the SoC toward an application. Third, measurement results of an implementation of the SoC in 40 nm CMOS will be given. The SoC achieves a 98.67% accuracy on a QRS-labeling task, while consuming only a total power consumption of 20.8 μW/Ch. and achieving 80 ms of latency.
- Finally, Chap. 6 will conclude the book by summarizing the main results.

Chapter 2
Literature Review

The previous chapter has given a broad overview of the field of wireless connected sensors, operating at the edge of the IoT. This book focuses on sensing devices for biomedical applications. The chapter will present a review of the literature on such systems. Note that the goal of this chapter is not to give a complete overview of all possible implementations of a certain building block, but rather to provide the reader a system-level insight in what each block entails.

In Sects. 2.1 to 2.6, the most important building blocks of a biomedical recording system will be revised, from the electrodes to the transmitter. For each building block, the most important design characteristics and trade-offs will be discussed, and pointers will be given toward the relevant works in the literature. In Sect. 2.7, a small case study of a wearable will be discussed, to highlight how the building blocks from the previous sections compare to each other in terms of power efficiency. From this case study, data transmission will appear as the main power bottleneck in such systems. In Sect. 2.8 several data compression and reduction techniques will be discussed, both in the digital and in the mixed-signal domain. Finally, Sect. 2.9 will summarize the key takeaways from the chapter and will formulate a conclusion from the literature review.

2.1 Overview of Biomedical Edge Sensing Devices

In recent years there has been a surge in research toward portable, wearable, and implantable medical systems [50, 71, 99, 159]. Due to an increasing and/or aging population, existing medical systems are put under pressure. In 2019, the average healthcare spending within OECD countries was 8.8% of the GDP, with costs ever rising since (partially also due to the COVID-19 pandemic) [96]. To counteract the rising medical costs, various solutions have been proposed for new monitoring and curative devices. One trend is to move toward preventive medicine, where the

J. Van Assche, G. Gielen, *Event-Driven Circuit Architectures for Scalable and Adaptive Sensor Readout*, Analog Circuits and Signal Processing,
https://doi.org/10.1007/978-3-031-97285-0_2

focus is on keeping the patient healthy and fit, rather than treating the patient after she/he gets sick [5]. To enable such continuous monitoring of healthy persons, the quality and reliability of the sensors in devices like fitness trackers, smartwatches, etc. should increase, such that the data coming from such devices become useful for medical diagnosis. Another trend is to monitor patients more outside of the doctor's cabinet or hospital (which is the current medical standard). For example, a condition such as epilepsy is difficult to monitor in a hospital setting and can be monitored for a long time in the home setting of a patient. Finally, recent trends also require the medical devices not only to monitor and collect data but also to autonomously treat the patient when needed. For example for seizures, an on-chip machine learning (ML) algorithm for seizure onset detection could be coupled back to a neurostimulator for seizure suppression [136].

To enable these new medical trends, technologies that continuously monitor physiological functions (like heartbeat, muscle contraction, brain activity, etc.) are needed. Such technologies should be able to monitor many vital signs of a person for long periods of time (days to weeks) and should be portable and non-obtrusive for the patient [50, 124]. This translates to a need for devices that work on a small battery (or no battery at all) and ideally are also wireless (as wires could limit a patient's mobility). With only limited energy available, the recording devices should consume as little power as possible. In Fig. 2.1, a high-level overview of such biomedical recording device is given. While various physiological signals can be measured on (or inside) the human body, many of these signals share a similar type of sensor architecture. This architecture includes the sensor readout—which consists of a low-noise amplifier (LNA), filters, and an analog-to-digital converter (ADC)—and a digital signal processing (DSP) block, a transmitter (TX), a power management unit, and reference voltages and bias currents. All of these subblocks are crucial for the functioning of the sensor system. This book focuses on the sensor readout and the DSP blocks.

2.2 Biosignals and Recording Artifacts

2.2.1 Biosignals

The human body and its organs function by means of many electro-chemical processes that are continuously repeated. As such, the human body can be seen as a source of electrical signals, which can tell a physician (or an algorithm) how a certain organ or bodily function is behaving. Examples of such signals are given in Fig. 2.1 and include the electrocardiogram (ECG), the electroencephalogram (EEG), and the electromyogram (EMG). The ECG is a potential that can be measured on the chest of a person and is caused by the repeated re- and depolarization of the muscle cells of the heart, which causes a rhythmic contraction (i.e., the heart beat). Due to the change in charge of these cells, an action potential (AP) can be measured outside

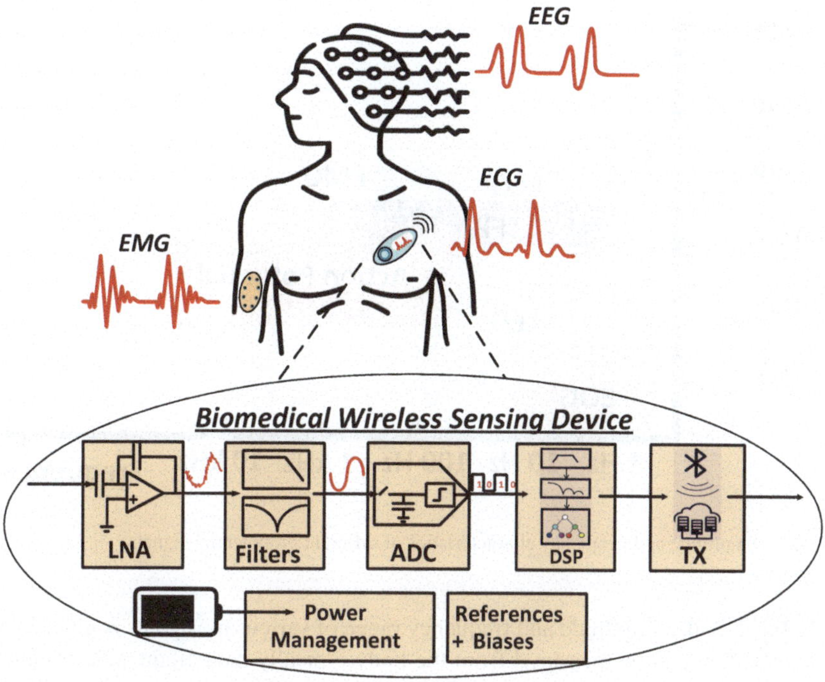

Fig. 2.1 Overview of a typical wearable scenario, in which (various) wirelessly connected sensor nodes are used. Indicated are the typical circuit building blocks of such system

the cells [50]. This potential is still measurable at the chest surface of a person. Since the ECG is directly linked to the functioning of the heart, it has become an indispensable tool for diagnostics in modern cardiology. Similarly to the ECG, the EMG measures the contraction of other muscles such as the biceps or triceps. These muscle cells exhibit a similar behavior compared to cardiac muscle cells, as they are also polarized when they contract, causing a measurable surface potential. A final example is the EEG, which is measured on the scalp of a person. The human brain functions by means of billions of neurons that communicate to each other via the axons. Neurons send electrical impulses (thus APs) via the axons, and the activity of these neurons is linked to many complex processes in the brain [50]. These APs can be measured by inserting an electrode in the brain [88]. They are not measurable individually outside the skull, as the spatial resolution of the neurons is too high and the amplitude of the APs would be too small to be measured on the outside due to the signal attenuation caused by the skull [50]. What is measurable at the scalp is the resulting potential of thousands of neurons that fire simultaneously. While no individual neuron activity can be distinguished from this signal, this resulting potential provides a neurologist information about a certain region of the brain such as the occipital lobe (vision), the cerebral cortex (motor functions), etc. The signal measured on the scalp is called the EEG [50].

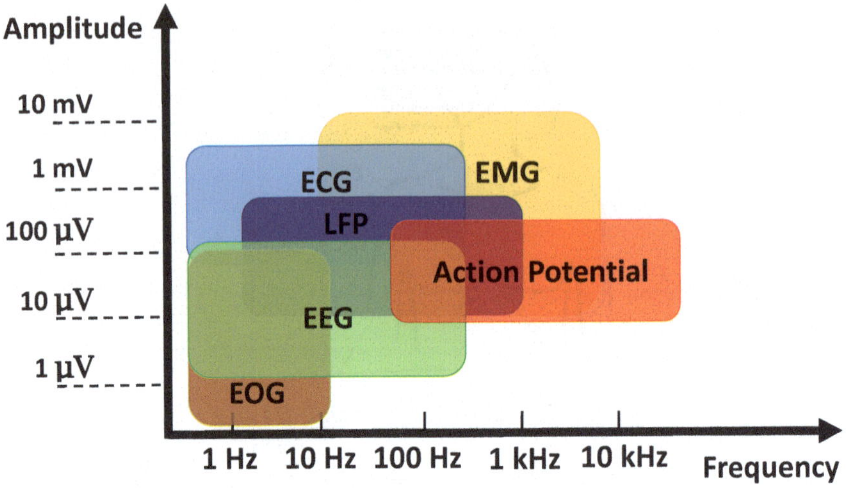

Fig. 2.2 Amplitude and frequency characteristics of different biopotential signals

In Fig. 2.2, the amplitude and frequency range of various biosignals is given. The biosignals that can be measured from the body typically have frequencies ranging from ≤ 1 Hz to ~ 15 kHz. From the plot it is also clear that these signals have only small amplitudes, ranging from 100 µV to a few mV. Hence, precise readout electronics are required to capture these signals. Note that, depending on the place where a certain signal is measured, the amplitude can vary. For example, the EEG signal measured on the scalp has a lower amplitude compared to the EEG signal that is measured inside the skull (which is also called i-EEG). However, the trade-off for this is the invasiveness of the recording methods, as recording from inside the body requires surgery to insert (part of) the electronics.

2.2.2 *Electrodes*

To measure biopotential signals, a transducer is required that provides a conductive path from the body to the measurement electronics, which is called the electrode [23, 25, 49, 180]. For any biomedical sensor circuit, the electrode is the first part of the signal chain. Hence, the type of electrode that is chosen for the application significantly influences the specifications of the subsequent readout circuits. The three main categories of electrodes for wearables and their equivalent circuit model are shown in Fig. 2.3: wet electrodes, dry electrodes, and noncontact (capacitive) electrodes.[1]

[1] Note that implantable devices such as neural recording arrays also use electrodes, and although these may not fall into the categories described in Fig. 2.3, they do provide a similar functionality.

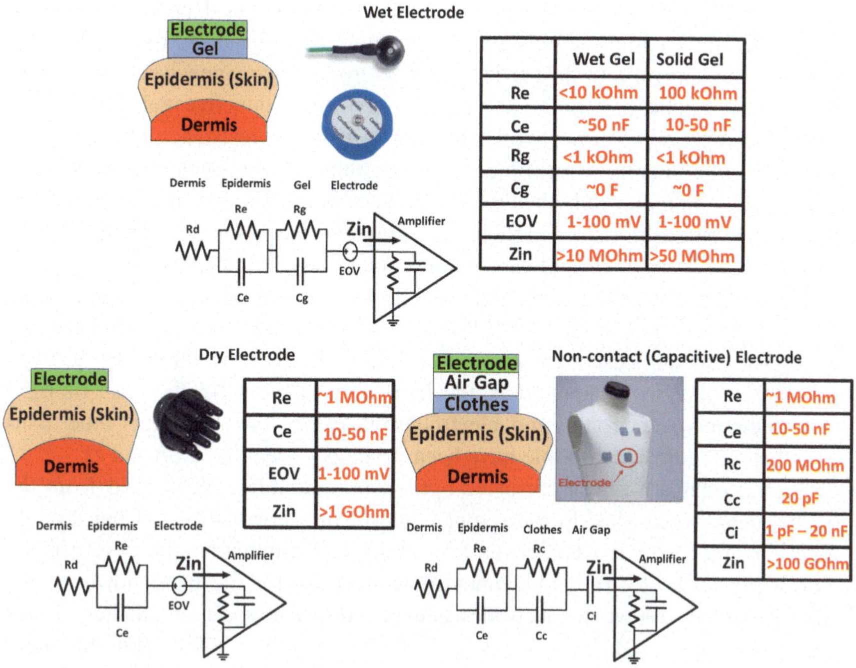

	Wet Gel	Solid Gel
Re	<10 kOhm	100 kOhm
Ce	~50 nF	10-50 nF
Rg	<1 kOhm	<1 kOhm
Cg	~0 F	~0 F
EOV	1-100 mV	1-100 mV
Zin	>10 MOhm	>50 MOhm

Re	~1 MOhm
Ce	10-50 nF
EOV	1-100 mV
Zin	>1 GOhm

Re	~1 MOhm
Ce	10-50 nF
Rc	200 MOhm
Cc	20 pF
Ci	1 pF – 20 nF
Zin	>100 GOhm

Fig. 2.3 Circuit equivalent model of various types of electrodes

- Wet electrodes are used in combination with a conductive gel that provides a low-impedance path [49]. An often used material for these electrodes is Ag/AgCl [180]. The wet electrode is till today the clinical reference in hospitals, and many wearable devices also use this type of electrodes. Due to the low electrode impedance and the conductive gel (which reduces the skin impedance), the total impedance can be around 10–100 kΩ in parallel with 10–50 nF [23, 25, 49, 180].[2] A low electrode and skin impedance is important because these impedances form a voltage divider with the input impedance of the readout electronics (see Fig. 2.3), which attenuates the biopotential signal. The lower the impedance of the electrode and the skin, the lower the input impedance of the input amplifier can be, which relaxes its circuit requirements (this will be discussed more in the next section). An input impedance of around 50 MΩ at 50/60 Hz would suffice for most wet-electrode applications [110].[3] Another advantage of a low electrode resistance is that the thermal noise that is generated by the electrode is fairly low,

[2] In literature, exact values for the impedance can differ, as this is also dependent on the size of the electrode, the gel that is used, the skin of the person on which the electrodes are attached, etc.

[3] The required input impedance also depends on the expected amplitude of the biosignal. For example, EEG has more stringent input impedance requirements compared to, for example, ECG [180], as the amplitude of the EEG signals is typically smaller.

as this noise is proportional to the resistance value [111]. Hence, this type of electrodes can capture the biosignals with a high signal-to-noise ratio (SNR).[4] Despite these favorable properties, wet electrodes face some limitations from a wearable application point of view. Firstly, the conductive gel that is used dries out after some time, which results in a higher resistance and degrades the recording quality over time. In a hospital setting, such electrodes can periodically be replaced by healthcare workers, but in a scenario where a person has to continuously use a wearable recording device, this is not practical [180]. Secondly, an important application requirement for wearables is that they need to be unobtrusive to the user: The wearable should seamlessly be integratable in the daily routines of its user [23]. Electrodes that use a wet gel can be found unhygienic or unpleasant by the user: Clothes can get dirty due to the gel, the gel might have to be applied in the hair of the person (in the case of EEG measurements), etc. To this end, electrodes that can be used for longer time spans and that are more comfortable have been developed (see Fig. 2.3b and c).

- Dry electrodes are a less obtrusive option for wearables. As is clear from the name, these electrodes do not require the use of a conductive gel but directly make contact with the epidermis (the skin). By eliminating the need for the gel, using such electrodes becomes more pleasant for the user; however, the skin resistance for such electrodes can be 10–100× higher than that of wet electrodes [49, 180]. This leads to a lower recording SNR, and the input impedance requirements of the readout circuits also become more stringent: An impedance of up to 1 GΩs can be required [23, 159]. Compared to wet electrodes, dry electrodes also have a more variable impedance. For wet electrodes, the conductive gel acts as an adhesive to the body; dry electrodes, on the other hand, rely on contact only (often, they are clamped to the body by means of a patch or headband [20]). This makes that such electrodes can move around when the user is moving [23], and hence, the electrode-skin connection is less stable. In the next subsection, it will be discussed how this affects the signal recording.
- A final type of electrodes is the noncontact (or capacitive) electrodes. This type of electrodes forms a capacitively coupled connection with the readout electronics and can be integrated in seating, clothing, headgear, etc. These electrodes are not only the most comfortable for the user but also the most challenging to use from an electronic readout perspective. The impedance of the clothes in which these electrodes are integrated can be 100× larger than the skin impedance [23, 26]. Hence, the required input impedance of the readout electronics can be up to 100 s of GΩs [23], which brings large design challenges. For example, the leakage current of the ESD diodes at the input of the recording application-specific integrated circuit (ASIC) can already limit the maximum achievable input impedance [23]. Moreover, these electrodes suffer from a high electrode noise and even larger impedance variations than dry-contact electrodes [23, 26].

[4] The real noise model for electrodes is more complicated than simply a thermal noise source; for a discussion on this the reader is referred to [25].

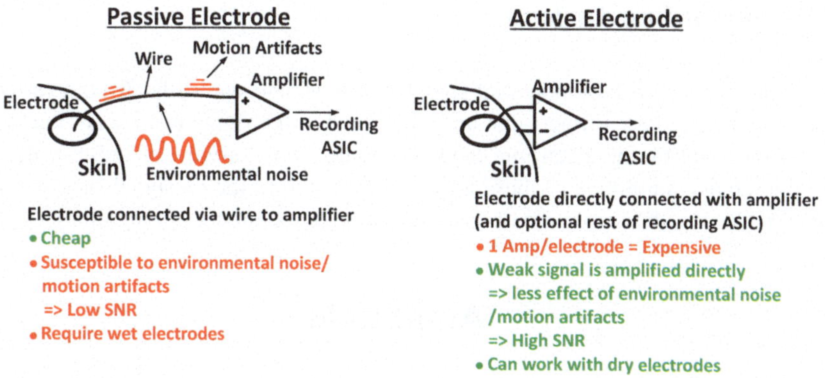

Fig. 2.4 Passive and active type of electrodes

From an application point of view, the type of electrode that is used is thus an important technological choice. Another important choice for an application is how the electrodes are connected to the readout electronics. For example, EEG electrodes are often distributed over the human scalp, to record signals from different areas of the brain [20]. The signals from all these distributed electrodes should eventually be connected to a central sensor node. In many cases, the electrodes are connected directly via wires to a central readout ASIC (see Fig. 2.4). This type of electrodes is called passive electrodes, as they do not involve any extra electronics [60, 180]. While this solution is simple (it only requires a single wire per electrode), if the wire to the readout ASIC is long, it can act as an antenna and pick up environmental noise (which originates from, e.g., the mains), which reduces the signal SNR. Moreover, cable movements may induce motion artifacts, which corrupt the signal recording [60, 180]. Therefore, passive electrodes should best be used with wet electrodes, which have a higher intrinsic SNR. Alternatively, active electrodes can be used (see Fig. 2.4). In this approach, each electrode has to be locally buffered by an integrated amplifier that directly amplifies the signal close to the electrode [20, 60, 180]. As the signal that is passed through the wire is much larger than the initial weak biosignal, the effect of environmental noise is largely mitigated. Moreover, since an amplifier is integrated close to the electrode, due to its low output impedance, this amplifier mitigates motion artifacts due to cable movements [180]. This approach is therefore also more compatible with dry and capacitive electrodes. A downside is the cost, as an amplifier (often a small ASIC) is required per electrode [20, 60, 180]. Each electrode also requires multiple cables (at least one ground, supply, and data cable per electrode). The active-electrode concept can be pushed further to also include a full wireless readout ASIC per electrode, each transmitting the data to a central base station [20]. While this approach is minimally obtrusive to the user, it does require a full ASIC + battery per electrode and, hence, is costly.

2.2.3 Recording Artifacts

When measuring weak biopotential signals, these signals can suffer from a variety of different recording artifacts that are not related to the signal of interest, but that do get recorded by the electronic readout circuits. An overview of these artifacts and the relative amplitude of these artifacts compared to the biosignals is given in Fig. 2.5. Clearly, the amplitude of these artifacts can be orders of magnitude larger

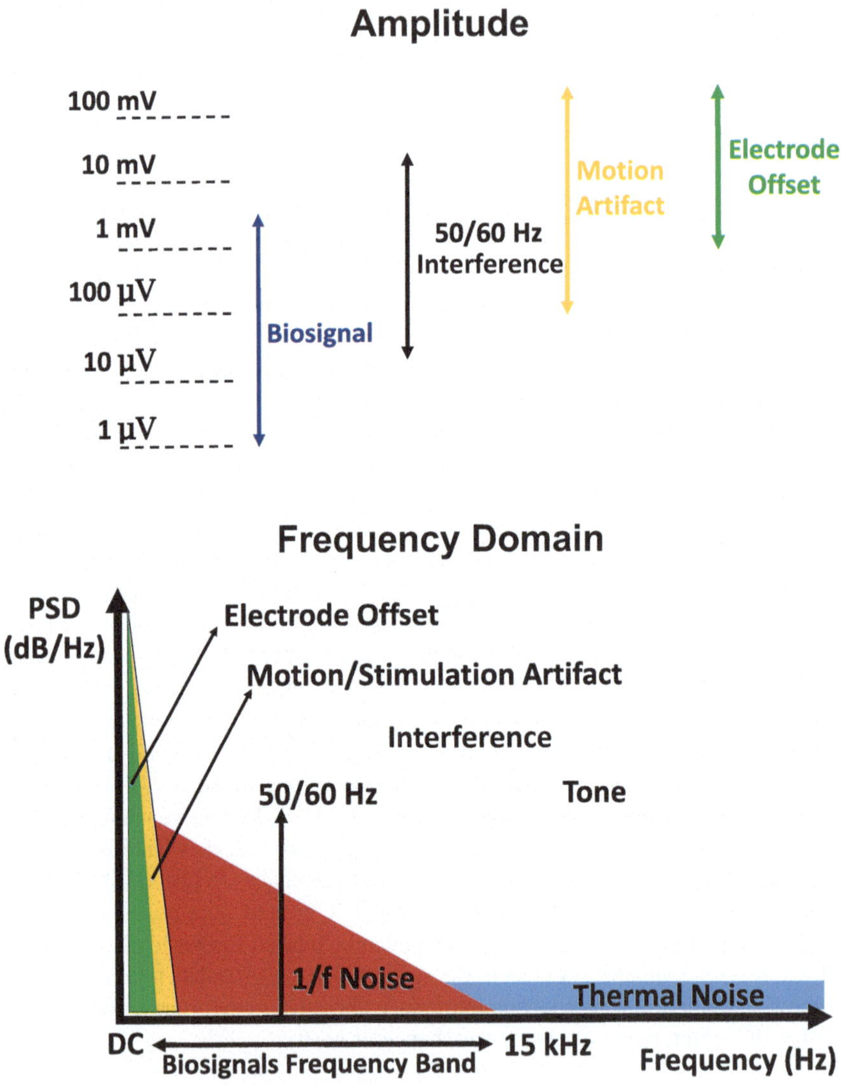

Fig. 2.5 Overview of the different signal artifacts and the typical order-of-magnitude amplitude (top) and frequency characteristics of different recording nonidealities (bottom)

compared to the biosignal's amplitude. Moreover, these artifacts overlap with the frequency band of the biosignals. Several different types of artifacts exist.

A first important artifact is the electrode offset voltage (EOV), see also Fig. 2.3, which is caused by the difference in electro-chemical reaction between the body (e.g., skin of the person) and the electrode (and if applicable gel) used for recording the signal. This offset can be up to 100 mV in amplitude [110, 159]. This artifact is (largely) a DC phenomenon, as depicted in Fig. 2.5. A second artifact to take into consideration is motion artifacts, which are caused by the impedance fluctuations that occur when using electrodes while a person is moving around. These motion artifacts are low-frequency ($<$ 20Hz) signals, correlated with the frequency of the movement that causes the impedance fluctuations (e.g., respiration, walking, etc.) [29, 159]. A third artifact is the presence of 50/60-Hz interference (also called power-line noise), which is caused by the main grid, that acts as a large transmitter antenna [60, 180].[5] These tones can be up to 10 s of mV in amplitude, and they overlap significantly with the signal frequency band of biosignals such as ECG and EEG.

Besides these recording artifacts, the electrodes and recording electronics also add noise to the signals. The two most important noise sources are 1/f noise (also called flicker noise) and thermal noise. 1/f noise has a power spectral density (PSD) that is inversely related to the frequency, and, while the origins are not fully understood, it is probably related to electronic trap states between two interfaces. For example, in NMOS and PMOS devices, the 1/f noise is most likely related to the traps in the gate oxide [105]. Thermal noise is the intrinsic noise that any current exhibits that flows through a conductor. The (voltage) noise PSD of, for example, a resistor is given by

$$v_n^2 = 4 \cdot k \cdot T \cdot R, \tag{2.1}$$

where k is the Boltzmann constant, T is the temperature, and R is the resistor value. Thermal noise is always present in any system and can thus only be reduced by using a smaller resistor (for lower voltage noise). This can also offer an (intuitive) explanation for the difference in intrinsic SNR between wet and dry electrodes [111].

2.3 Analog Front-End: Low-Noise Amplifier and Filters

This section will discuss the LNA and filters used in wearable biomedical systems. Together, these circuits are referred to as the analog front-end (AFE).

[5] Note that in some measurement laboratories, this prompts researchers to do measurements without the lights on, as the (switching) power electronics of the lights are believed to be the main cause of this interference. The measurement equipment is, however, also connected to the mains, and interference may couple to the readout electronics through this way; thus it is doubtful that turning off the lights makes any difference on the recording quality whatsoever.

2.3.1 Low-Noise Amplifier

The previous section has highlighted how biomedical sensor interfaces have to capture low-bandwidth signals with amplitudes that only range from 100 μV to several mV. To capture such small signals, an LNA, also called an instrumentation amplifier (IA), is used. Its primary function is to amplify the weak biosignals, such that the subsequent circuits can process the signal more easily, without having to design all circuits for extremely low noise levels. However, this comes at a cost. The input-referred noise (IRN) of the LNA should be small enough compared to the signal of interest that is recorded, in order not to corrupt the sensor measurement. Looking initially only at thermal noise, the noise requirement of the LNA is directly related to the power consumption of the LNA. Since the LNA consists of transistors, the thermal noise of the transistors adds to the total amount of noise of the amplifier circuit. The thermal noise PSD of a single transistor is given by

$$v_n^2 = \frac{4 \cdot k \cdot T \cdot \gamma}{\text{gm}}, \tag{2.2}$$

where gm is the transistor's transconductance and γ a factor to take into account small-channel-length effects (assume it is equal to 2/3 for long-channel transistors). As the transconductance is proportional to the bias current (and hence the power consumption) of the transistor, it becomes obvious that, to achieve lower noise, a penalty in power consumption has to be paid. For this reason, a figure of merit (FoM) has been introduced, called the noise efficiency factor (NEF), which quantifies the trade-off between the bias current of an amplifier and its IRN [43, 143]. The NEF is a unitless quantity, which is defined relative to the IRN of an amplifier with bipolar devices.[6] The NEF for an amplifier is given by

$$NEF = v_{\text{n,rms}} \cdot \sqrt{\frac{2 \cdot I_{\text{tot}}}{4 \cdot V_{\text{T}} \cdot k \cdot T \cdot f_{\text{BW,LNA}} \cdot \pi}}, \tag{2.3}$$

where $v_{\text{n,rms}}$ is the total integrated IRN of the amplifier, I_{tot} is the total bias current that the LNA consumes, V_{T} is the thermal voltage (which is equal to $\frac{k \cdot T}{q}$, where q is the charge of an electron), and $f_{\text{BW,LNA}}$ is the amplifier bandwidth. Related to the NEF is the power efficiency figure (PEF), which is given by

$$PEF = \text{Vdd}_{\text{amp}} \cdot NEF^2 \tag{2.4}$$

with Vdd_{amp} the amplifier's supply voltage. The PEF gives a more complete picture of the noise-power trade-off, as also the supply voltage is taken into account. The

[6] Bipolar amplifiers were believed to achieve the best noise efficiency, which is true compared to a MOS device.

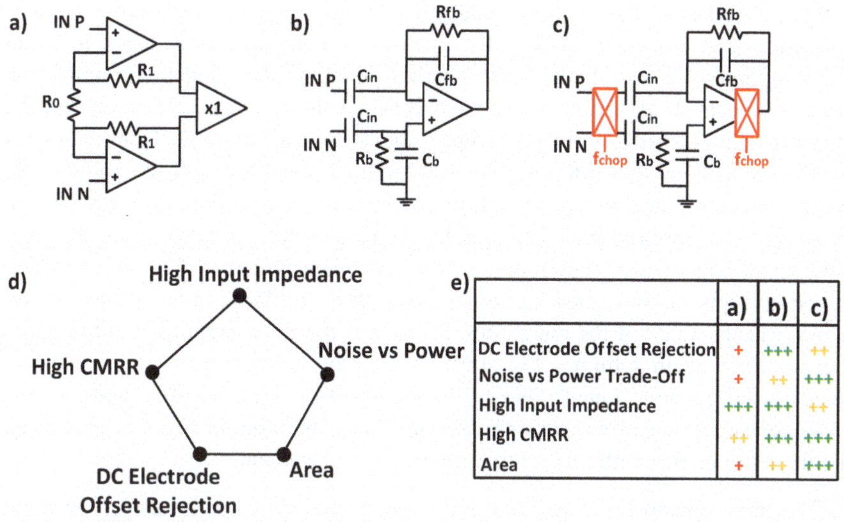

	a)	b)	c)
DC Electrode Offset Rejection	+	+++	++
Noise vs Power Trade-Off	+	++	+++
High Input Impedance	+++	+++	++
High CMRR	++	+++	+++
Area	+	++	+++

Fig. 2.6 Circuit overview of various LNA topologies: (**a**) the three-opamp IA, (**b**) the capacitively coupled IA, and (**c**) the CCIA. Given are also (**d**) the different design requirements for the instrumentation amplifiers and (**e**) the table with a performance comparison of the three topologies

NEF/PEF FoM values thus provide a metric for the quality of a designed LNA circuit. For a MOSFET-based amplifier, the minimum NEF that can be theoretically achieved is 2.02 [43].[7] In practical implementations, designs consume more power due to, e.g., biasing networks. Moreover, the NEF assumes that the thermal noise is the dominant noise source of the amplifier. However, as seen in the previous section, 1/f noise is also an important noise source for low-frequency signals. Mitigating 1/f noise can cause excess power consumption. Hence, it can lead to NEF values that are higher than theoretically predicted. Lastly, like all FoMs, the NEF fails to capture all the design trade-offs that a circuit designer has to make.

Figure 2.6 shows three popular LNA topologies: the three-opamp IA, the capacitively coupled IA, and the capacitively coupled chopper-stabilized IA (CCIA). These topologies are all (pseudo-)differential amplifiers that require two measurement electrodes. Either both electrodes contain a signal (hence, the LNA works fully differentially) or what is done more often for wearables, the measurement is done between a signal electrode and a reference electrode [29, 47, 88, 95, 159, 191]. This reference electrode is placed away from the measurement electrode, to provide a clean reference voltage that does not contain the input signal.

[7] Over the years, many innovations in LNA design have led to designs that surpass this "theoretical" minimum. A full discussion of these amplifier topologies is outside the scope of this work, but the interested reader is referred to [43] for a comprehensive overview.

Figure 2.6d also shows the different design requirements for LNAs. Besides low noise and power, the LNA also requires a high input impedance to avoid signal attenuation (see the previous section on electrodes) and to minimize the effect of electrode impedance mismatch between the reference electrode and the measurement electrode [95]. Interference signals such as 50/60 Hz from the mains or motion artifacts will often appear as a large common-mode (CM) signal [23]. Hence, the amplifier also requires a high common-mode rejection ratio (CMRR), to avoid that the amplifier gets saturated by the CM signal. As LNAs typically only have small bandwidths, they require a long recovery period if they get saturated, which corrupts the biosignal measurements [191]. Similarly, the amplifier should also reject the EOV of the electrodes, because if the EOV amplitude is too large, the amplifier again saturates [34, 159]. A solution to these problems can be to increase the dynamic range (DR) of the readout front-end, but this increases the power consumption of the sensing system [110]. The table in Fig. 2.6e compares the three topologies with respect to each design requirement.

- The three-opamp IA (Fig. 2.6a) is a popular choice of IA. It has a high input impedance since the input of the opamp is the gate of a transistor [1]. However, it does not score well on the other requirements compared to the other two topologies. The CMRR of the amplifier depends on the matching of the resistors [1, 159]. Moreover, the circuit has to use three opamps, and the resistors add extra noise, on top of the noise of the input opamps, which leads to an unfavorable noise-power trade-off [159]. Using multiple opamps also makes that the circuit has a large area, compared to the other topologies that use a single opamp. Lastly, the topology is directly coupled (also called DC-coupled) to the electrode, which makes that the offset is not rejected, and hence, is amplified together with the signal [159]. To avoid saturation, the gain of the amplifier is to be limited, which increases the noise requirements for the following blocks.
- The capacitively coupled IA (Fig. 2.6b) is an alternative to the three-opamp topology [47]. Due to the AC-coupled capacitors at the input, this amplifier provides an infinite input impedance at DC. Moreover, these input capacitors form a high-pass filter with the feedback resistor and capacitor. This filter effectively cancels the EOV of the electrodes [47]. A high CMRR can also be achieved, but care has to be taken that the input capacitors are well matched. This can be done by using large external capacitors or by employing trimming capacitors [158]. The capacitively coupled amplifier thus seems a promising solution for an amplifier; however, its area and noise-power trade-off are limited by the 1/f noise of the transistors. To mitigate the 1/f noise, the device sizes of the amplifier's transistors should be upsized, as the 1/f noise is proportional to the inverse of the transistor area ($\sim \frac{1}{W.L}$). Upsizing the devices leads, however, to excess power consumption and a larger circuit area [18].
- Since the 1/f noise limits the capacitively coupled IA topology in area and power consumption, many works in literature nowadays use the CCIA topology (Fig. 2.6c) [18]. This amplifier introduces chopping switches before the capacitors, which mitigate the 1/f noise and offset of the amplifier. As this amplifier

allows to design the amplifier close to the thermal noise limit, it achieves a low power consumption, and the device sizes can be chosen small, leading to a compact LNA design [18, 34]. Moreover, a high CMRR can be achieved by implementing dedicated CM cancelation circuits [18]. The CCIA seems to offer many advantages; however, introducing chopping also comes with its challenges. Chopping degrades the offset rejection properties of the capacitors, and hence additional circuits are required to compensate the offset [34]. It also degrades the input impedance of the amplifier, since the chopping switches together with the capacitors form a switched-capacitor resistor at the input of the LNA [18]. Nonetheless, the CCIA remains an attractive solution for low-power and compact LNA implementations. This topology is therefore discussed further in the following paragraphs.

In Fig. 2.7, the operating principle of a chopping amplifier is depicted. Chopping is employed to remove low-frequency 1/f noise and DC offsets of an amplifier

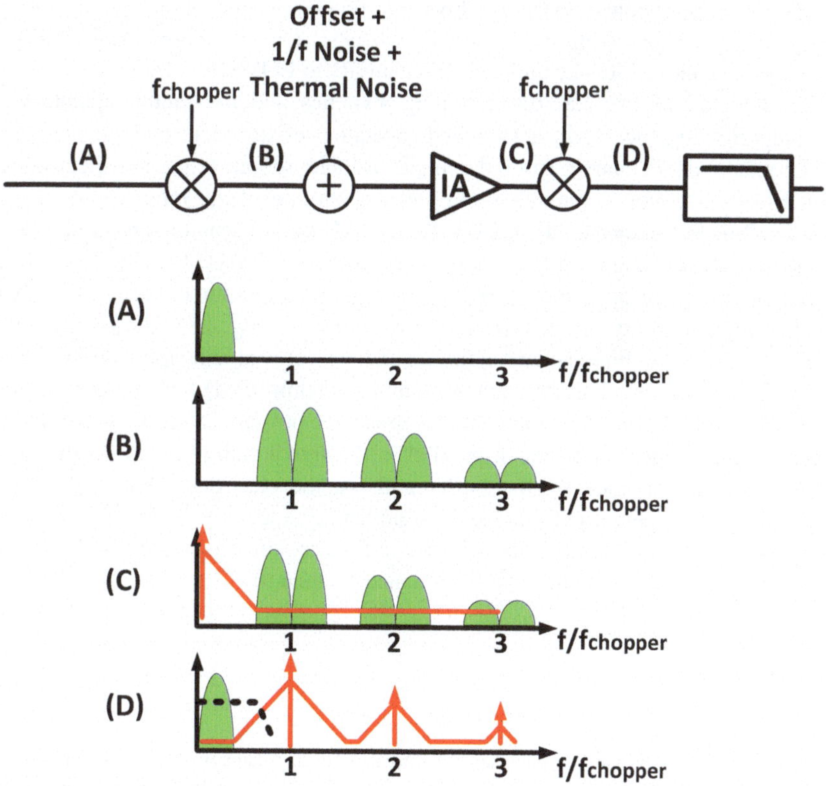

Fig. 2.7 Operating principle of a chopping amplifier

[18, 34].[8] A chopper switch is equivalent to a mixer, which multiplies the incoming signal with the chopper signal (a square wave). This up-converts the input signal (e.g., the biopotential) from baseband to $f_{chopper}$ (see Fig. 2.7b). This frequency should be chosen such that the signal bandwidth is spaced far enough from the "1/f-corner" frequency, which is the frequency above which the thermal noise PSD starts to become dominant over the 1/f-noise PSD. Once the signal is up-modulated, the high-frequency signal passes through the IA, which adds thermal noise, a DC offset, and 1/f noise. Note that the bandwidth of the IA should be sufficiently high, such that the up-converted signal is not distorted [18, 34]. After the signal passes the IA, the signal is amplified and contains the 1/f noise and offset at baseband (Fig. 2.7c). Subsequently, the signal passes again through chopping switches at the output of the IA, which up-modulates the 1/f noise and offset but down-modulates the original input signal. This results in the final output given in Fig. 2.7d, where the signal is amplified, and the offset and 1/f noise are modulated out of the signal band, where they can be easily filtered out with a low-pass filter [18, 34].

The chopping amplifier thus offers properties that match well with the requirements for a biomedical system. However, related to the chopping operation, the topology suffers from two main drawbacks: a lower (chopping-frequency-dependent) input impedance and a high sensitivity to EOV [18, 34].

As depicted in Fig. 2.8, the chopping switches and the input capacitors of the amplifier together form a switched-capacitor resistor. The chopper switches continuously draw charge from the input voltage source (e.g., the electrodes), which leads to a continuous charging and discharging of the input capacitor C_{in} [17, 18, 34]. For example, a CCIA with a 10-pF input capacitor and a chopping frequency of 20 kHz has only an input impedance of 2.5 MΩ [17, 18]. Two circuit techniques are therefore typically used to increase the input impedance of the LNA: precharging amplifiers and positive feedback loops (see Fig. 2.8). These techniques suffice to satisfy the impedance requirements for most wearable applications.[9]

In Fig. 2.8, the precharging technique and the (simplified) timing are depicted. Each cycle before the chopping signal switches polarity, a duty-cycled buffer precharges the input capacitor, in parallel to the signal path [17, 18]. In this way, most charge is provided by the buffer and not the electrode. Hence, the input impedance of the signal path increases significantly, ideally to infinity. Due to the finite buffer gain and bandwidth, however, a nonzero error is made that limits the impedance. The approximate input impedance is equal to [17]:

[8] Another popular technique to compensate 1/f noise and offsets is auto-zeroing. However, it is less used in biopotential amplifiers, as it involves a two-phase sampling operation, which leads to noise folding into the signal band [180].

[9] Some applications with noncontact electrodes such as [23, 26] require even larger input impedance values which can likely not be obtained by these techniques. For such applications, another IA topology such as the capacitively coupled IA could be used.

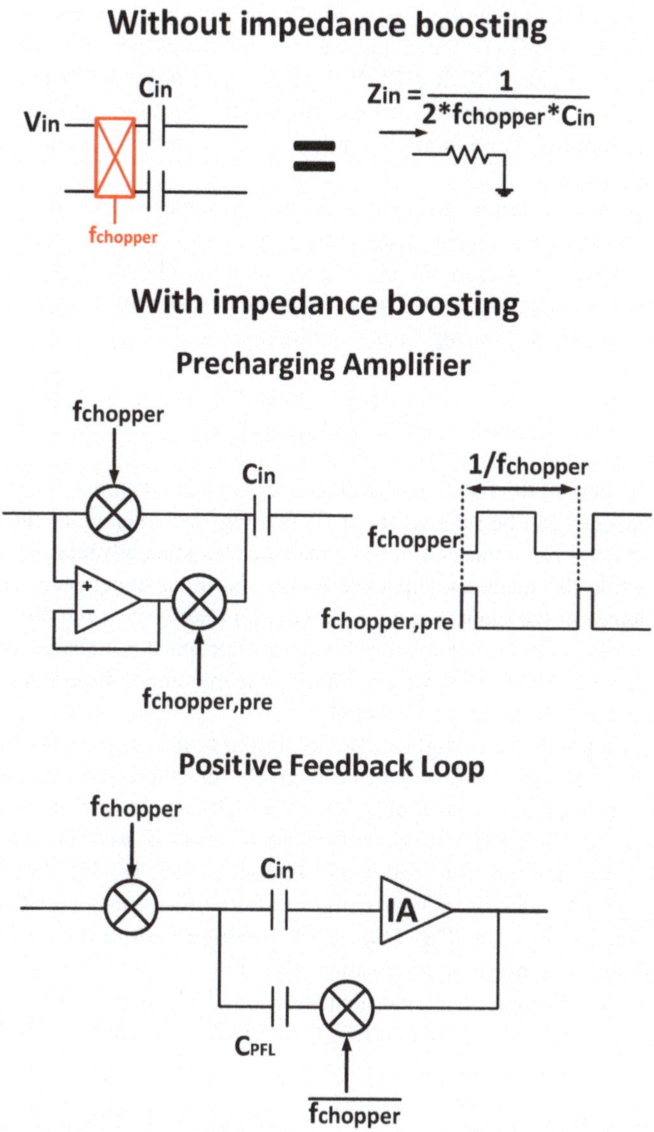

Fig. 2.8 Impedance boosting circuits for chopper-stabilized amplifiers: precharging amplifier and positive feedback loop

$$Z_{\text{in}} = \frac{1}{2 \cdot f_{\text{chopper}} \cdot C_{\text{in}}} \cdot \frac{\exp(T_1/\tau)}{1 + \beta \cdot \exp(T_1/\tau)}, \quad (2.5)$$

where β is the gain error of the buffer, T_1 is the precharge time of the buffer, and τ is the time constant of the precharge buffer. The precharging technique allows to significantly boost the input impedance. In [18] an impedance increase of over 75× could be achieved. However, the technique also introduces extra noise to the signal path, leading to a reduced SNR [17]. If there is a gain mismatch between the two buffers, there is also a mismatch in input impedance between the two differential inputs of the amplifier. For a more in-depth discussion on this technique, the reader is referred to [17, 18].

Another popular technique to boost the input impedance of IA is using a positive feedback loop [34]. In this technique, charge is inserted in each chopping period by means of a feedback loop from the IA output back to the chopper input by means of the capacitor C_{PFL} (see Fig. 2.8). Theoretically, again, an infinite input impedance can be achieved, when choosing C_{PFL} as follows:

$$C_{PFL} = \frac{C_{in}}{A_{v,IA} - 1} \tag{2.6}$$

with $A_{v,IA}$ the gain of the IA. Note that if a high gain is used and/or C_{in} is small, the feedback capacitor can be relative small. This brings a challenge, as the mismatch from this capacitor can undo part of the impedance boosting advantage [17, 18, 34]. Moreover, while the feedback capacitor boosts the input impedance, the bottom-plate capacitance of the input capacitor C_{in} ($C_{in,bp}$) also forms a switched-capacitor resistor with the chopping switches. This parasitic capacitor's charge draw is not compensated for by the feedback loop, limiting the maximum impedance boost that the technique can achieve (e.g., 5× in [34]).

The CCIA topology, at first glance, looks similar to the capacitively coupled IA. However, since the chopping switches effectively turn the input capacitors into a resistor, the input capacitor no longer acts as a high-pass filter. Since the EOV of electrodes can be fairly large, if no compensation circuit is used, the maximum IA gain before saturating the front-end would be limited [17, 18, 34]. To compensate for this, many CCIA topologies form a high-pass filter by means of a DC servo loop (DSL), depicted in Fig. 2.9. This DSL can be realized both in the analog domain [17, 18, 34] and in a mixed-signal fashion [91]. The DSL integrates the IA output and extracts the DC component and then feeds this DC component back to the IA input to create a high-pass corner (for ECG or EEG, this corner frequency is often

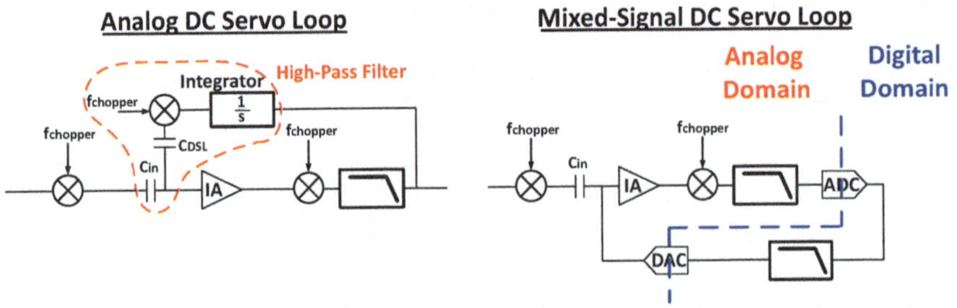

Fig. 2.9 Possible DSL implementations for electrode offset compensation

taken around 0.5 Hz [180]). For example, in the analog implementation of the DSL in Fig. 2.9, an integrator circuit is connected to the output of the IA, and the output of the integrator is up-converted and passed to the virtual ground of the IA by means of a capacitor C_{DSL}. This loop creates a high-pass filter with the following corner frequency [34]:

$$f_{\mathrm{c,HPF}} = \frac{C_{\mathrm{DSL}}}{C_{\mathrm{fb}}} \cdot f_{\mathrm{u}}, \tag{2.7}$$

where C_{fb} is the feedback capacitor of the CCIA, and f_{u} is the unity gain bandwidth of the integrator. The DSL in the analog domain can cancel the EOV completely, but this comes at the cost of a large circuit area per IA (which is also not able to be multiplexed). The mixed-signal DSL solves this problem, by implementing the integrator in the digital domain and applying offset compensation by means of a digital-to-analog converter (DAC). However, due to mismatch/resolution limitations in the DAC, the DC offset is not fully canceled [91].

As a final note on the LNA, the driven right leg (DRL) circuit is discussed (Fig. 2.10). All circuits of the recording ASIC share a ground, but this ground is not connected to the body it is recording from. As the human body is electrically floating compared to the ASIC, it acts as an antenna and picks up disturbances from, e.g., the mains, and hence, large CM signals can be present at the input of the IA. Therefore, to improve the CMRR of the IA, a circuit called DRL is used [141]. This circuit extracts the CM voltage of the amplifier from the IA output and drives the potential on the human body (via a reference electrode) to this CM voltage,[10] hence decreasing the amplitude of the external interference. Moreover, for DC-coupled circuits such as the three-opamp IA, it provides a stable biasing point at the input of the LNA.

2.3.2 Filters

After the IA, the AFE typically includes an additional filter, to remove any unwanted signal components that are not filtered out by the IA. An overview of various filtering techniques for various artifacts, such as 50/60-Hz noise, out-of-band interference, and motion artifacts, is given in Fig. 2.11. As indicated in the figure, filters can be implemented in the analog, mixed-signal, or digital domain [11, 19, 29, 52, 59, 107, 108, 158, 159, 173].

Analog filters can be implemented as simple passive filters or active filters. Some implementations of active filters include the Gm-C type of filters, active RC filters, etc. Some recent designs from literature can be found in [19, 52, 107, 173]. Mixed-signal filters consist of a feedback loop between the DSP which implements a

[10] As the name suggests, the convention is to attach this reference electrode to the right leg of the person (as is, e.g., done for ECG measurements in the hospital). Note that this can be done on any part of the body.

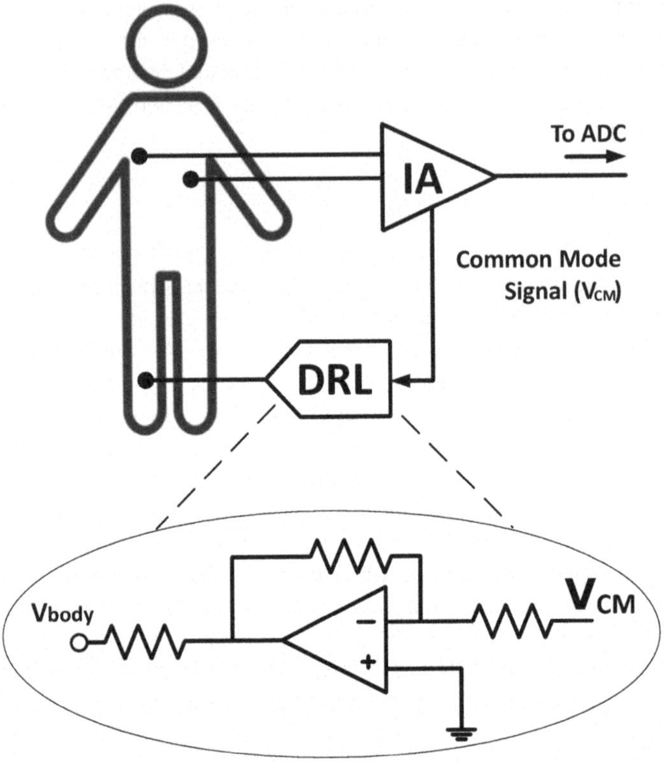

Fig. 2.10 DRL circuit used in wearable biomedical systems

filter that filters the incoming ADC data, and a feedback DAC, that takes care of filtering out the artifacts (similar to a mixed-signal DSL) [11, 158, 159]. Digital filters implement the filters completely in the DSP domain and can consist of infinite impulse response (IIR) or finite impulse response (FIR) filters (see Fig. 2.11), transform algorithms such as, e.g., the wavelet transform or blind source separation algorithms such as independent component analysis or principal component analysis [7, 59]. Implementations in each domain hold advantages and disadvantages, which are summarized in Fig. 2.11. A detailed implementation of each possible filter type would be out of scope for this work.

Rather, for a few signal artifacts that can occur in a wearable system, some possible filter techniques will be discussed.[11]

[11] Unlike the IA, which practically any wearable system uses, not every system needs all the different types of filters that will be discussed in this subsection. Depending on the application, some filters can be omitted, because the artifacts are less outspoken in those applications. For example, motion artifacts are less present when a system uses wet electrodes or when the system is not intended for ambulatory usage.

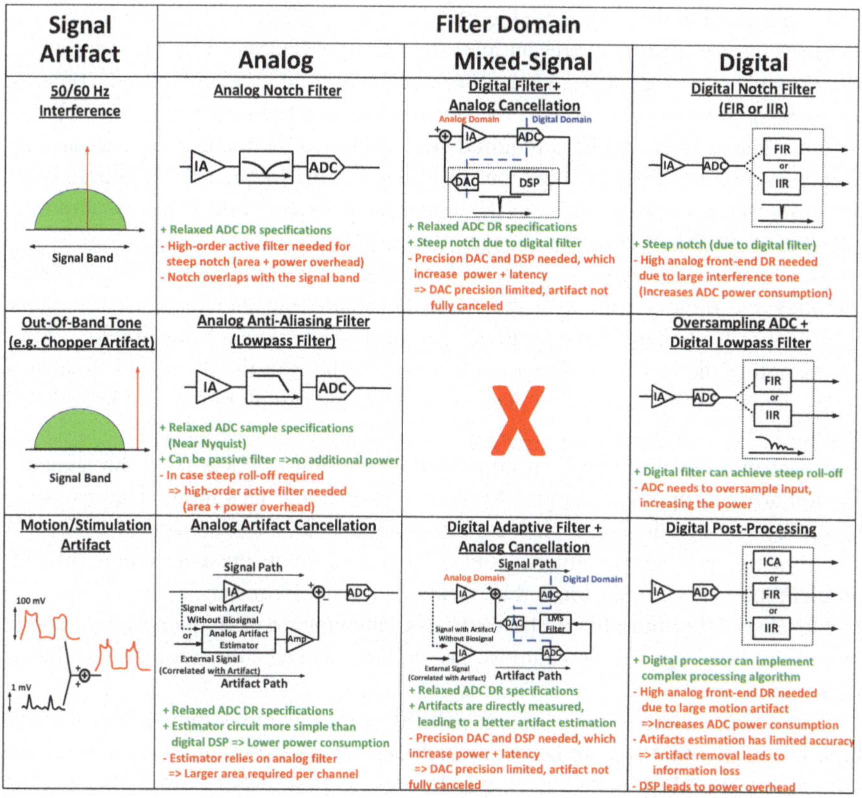

Fig. 2.11 Overview of various filter types used in biomedical systems and the artifacts they cancel

Power-line noise/interference, e.g., 50/60-Hz tones, appears as large common-mode signals at the IA input. Due to the high CMRR, they are partially filtered out by the IA, but some residual artifacts remain present after the IA.[12] To remove the power-line noise, typically a notch filter is employed [11, 19]. This type of filter is a narrowband bandstop filter, where the center frequency is placed at 50/60 Hz. The downside of using notch filters is that they overlap with the frequency band of biosignals such as ECG or EEG, and hence, some signal information is lost when applying such a filter. Hence, a high Q-factor is required to minimize the information loss. A digital or mixed-signal filter is therefore often a better option to use, because they can achieve a higher Q-factor compared to an analog implementation, which would require a power-hungry, high-order filter.

[12] In measurement setups for chips, these signals also show up due to coupling through the supply and ground lines. Hence, for measurements where a very pure spectrum needs to be measured, sometimes battery-powered setups are used.

As discussed in the section on the IA, if chopping is employed in the IA, an out-of-band chopper artifact is present after the IA, which needs to be filtered out [17, 18]. Possible other out-of-band signals (e.g., RF interference) might also be present. To avoid such tones aliasing into the signal band (see the next section on ADCs), an anti-aliasing filter (AAF) is required. The AAF in a biomedical system can be a simple passive RC filter [17, 18], but if a steep cut-off frequency is required (e.g., if the interference tone is close to the signal band), active filters may be required, which increase the AFE power consumption significantly [107, 173].

Alternatively, the AAF can be implemented in the digital domain, by over-sampling the input signal with the ADC. Oversampling increases the frequency beyond which aliasing occurs but increases the ADC power consumption. After oversampling the signal, a digital filter can later filter out the unwanted frequency components, and by means of a decimation filter the sampling rate can be reduced again.[13]

A last type of filter can be used to remove motion artifacts from the signal.[14] Motion artifacts appear as large CM input signals at the IA's input. They are low-frequency AC signals, but, unlike, e.g., power-line noise, the characteristic signal is not known a priori. Hence, filtering out the motion artifacts presents a more difficult problem than interference tones that occur at a known frequency.

Moreover, over time the motion artifacts may change in frequency, e.g., a user in rest, mildly jogging or running has motion artifacts that occur at different frequencies.

A first approach to remove motion artifacts is a fully digital approach, in which a high-DR AFE and ADC are needed, which can sample both the biosignals and the motion artifacts [110]. This increase in DR results in a higher power consumption for both the AFE and the ADC. After sampling, digital filtering methods try to recover the biosignal from the corrupted signal with motion artifacts. Several algorithms exist for this. The simplest method is to use high-pass filters, assuming that the motion artifact is low frequent in nature. However, there is a significant signal loss with this method, since the low-frequency information of the signals is also filtered out. More elaborate methods include blind source separation methods such as principal component analysis or independent component analysis, which decompose biosignals into statistically independent components that are associated with both the motion artifact and the signal [7, 59]. By removing components related to the motion artifact, the artifact is effectively canceled. Similarly, the wavelet-based transform method applies a transformation to the signal and then removes the coefficients related to the signal. This results in a residue that only contains

[13] Note that no mixed-signal implementations exist, as aliasing is caused exactly by the ADC or DAC operation.

[14] For closed-loop neurostimulators, a similar type of artifact, the stimulation artifact, exists in a similar fashion [17, 18]. One significant difference though is that stimulation artifacts are caused by the electronics themselves, unlike motion artifacts that are caused by external factors, and can be anticipated for in the design of the system [15].

the motion artifact, which is subsequently subtracted from the original signal to obtain a clean output signal [7]. However, both the blind source separation and transform-based methods assume that the motion artifacts and the signal are easy to separate. When the motion artifact or the signal (like, e.g., ECG) is periodic in nature, this can be the case, but when the signal or artifact is nonstationary (like, e.g., an EEG), these approaches may fail [29]. Moreover, these methods require a high DSP power consumption to implement the independent component analysis or wavelet transform [7, 29, 59].

To reduce the DR requirements of the AFE and ADC, mixed-signal approaches based on adaptive filtering have been proposed [130, 158, 159]. These approaches try to estimate the motion artifact by using a signal that is correlated to the motion artifact, but that does not contain the biosignal. This can, for example, be an external accelerometer, or a bio-impedance measurement can be performed in the same readout channel as the biosignal [130, 158, 159]. Mixed-signal approaches digitize the signal correlated to the motion artifact, and by means of an adaptive filter (e.g., a least-mean-square filter), the artifact is canceled in the signal path by a DAC. This reduces the DR requirements of the ADC significantly. Moreover, these techniques directly estimate the motion artifact, which avoids relying on statistical properties of the signal like fully digital techniques. However, due to the limited precision that can be achieved by the DAC due to power/area constraints and due to the low latency requirements of the adaptive filter, which limits the filter's complexity, mixed-signal approaches do not fully cancel the artifact [11, 91, 158, 159]. Moreover, the adaptive filter requires parameter tuning of the filter coefficients, which depend on the specific use case in which the wearable is used [29]. This can limit its usability in practical applications.

A last approach is a fully analog cancelation of motion artifacts [29, 108]. This approach avoids the use of power-hungry DSP and/or the need for high-resolution DACs. Similar to the adaptive filtering, the motion artifacts are estimated, with an analog filter (see Fig. 2.11), and directly subtracted from the corrupted signal in the signal path [29, 108]. For example, in [29] for a setup with capacitive electrodes, the motion artifacts are estimated by a second capacitive electrode, which does not contain the input signal, but which is subject to the same impedance variations as the signal electrode. As both signals are in the analog domain, ideally the artifacts can perfectly be canceled (however, a slight increase in noise floor can be observed). A disadvantage of canceling the artifacts in the analog domain is that it has to be implemented per recording channel and that it cannot be multiplexed easily.

2.4 Analog-to-Digital Converter

The next building block of the biomedical system is the ADC, which converts the amplified and filtered output of the AFE into a digital signal. This signal can then

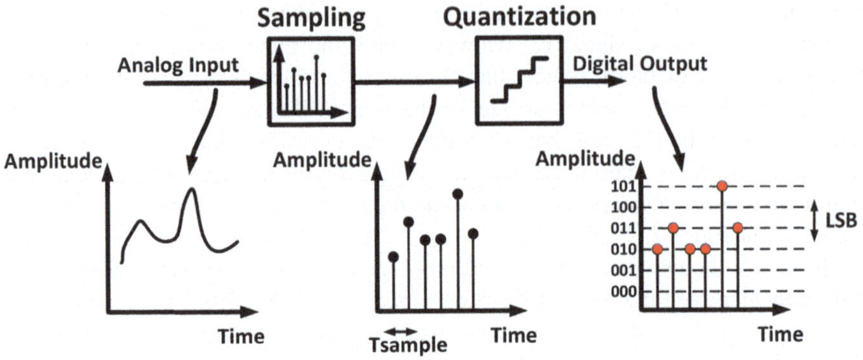

Fig. 2.12 Block diagram of an ideal ADC. Highlighted are the two fundamental processes: sampling and quantization (top). Indicated are also the different time-domain waveforms in each part of the conversion process (bottom)

subsequently be processed by the DSP. Below, first some ADC basics are revised, before discussing some often used ADC topologies and performance metrics.[15]

2.4.1 *The Ideal ADC*

For an analog-to-digital conversion, two fundamental processes are performed: sampling and quantization. These two processes and the resulting time-domain waveforms are depicted in Fig. 2.12. As depicted, sampling involves the periodic evaluation of the continuous-time (CT) input, which results in a time-discrete output signal with continuous amplitude. The quantization process then turns this CT amplitude into a discrete amplitude value, e.g., a digital code. In both processes, information about the input signal is lost. Hence, depending on the required specifications of the ADC, a higher sampling rate/finer quantization is required. Important to note is that these processes can be interchanged: Many ADCs sample first and then quantize these samples, but, for example, CT ADC architectures (e.g., CT pipeline ADCs, CT delta-sigma ADCs, level-crossing ADCs, etc.) perform (part of) the quantization first, before sampling the signal. Sampling and quantization are now discussed in more detail.

[15] Note that while this book focuses on biomedical systems, the discussion on ADCs is generally applicable to any application that requires an ADC. This section might be a bit more elaborate compared to the other sections, as ADCs will become more important in the following chapters of this book.

2.4.1.1 Sampling

Most modern DSPs work in discrete time (DT); hence, operations are performed one clock cycle at a time. Thus, to interface the DSP with a CT analog signal, a time discretization needs to occur, which is done by the sampling process. In this sampling process, the analog signal is evaluated every sample period T_{Sample}. In between sample periods, the sample process discards the CT input; this information is lost in the sampling process. In the time domain, this sampling process can be described mathematically by means of the Dirac function $\delta(t)$. The periodic evaluation of a signal is equivalent to multiplying the input signal with a train of Dirac pulses that are spaced by T_{Sample}. The sampled signal in the time domain can thus be written as

$$V_{\text{s}}(t) = V(t) \cdot \sum_{n=-\infty}^{+\infty} \delta(t - nT_{\text{Sample}}) = \sum_{n=-\infty}^{+\infty} V(nT_{\text{Sample}}), \tag{2.8}$$

where $V(t)$ is the CT analog input and $\sum_{n=-\infty}^{+\infty} \delta(t - nT_{\text{Sample}})$ is the train of Dirac pulses. As is known from signal theory [105], a time-domain multiplication is equivalent to a convolution in the frequency domain. Therefore, the inverse of this sample period, the sampling frequency $f_{\text{Sample}}(= \frac{1}{T_{\text{Sample}}})$, can also be used to characterize the ADC sampling operation. The sampling operation in the frequency domain is given in Fig. 2.13. Applying the Fourier transform to the sampled time-domain signal then yields the corresponding frequency-domain spectrum[16]:

$$V_{\text{s}}(f) = \frac{1}{T_{\text{Sample}}} \cdot \sum_{n=-\infty}^{+\infty} V(f - nf_{\text{Sample}}). \tag{2.9}$$

Figure 2.13 depicts the frequency spectrum for various sampling frequencies. A sampled signal will have a spectrum that has a baseband signal that is equal to the original and replicas, also called alias bands, around integer multiples of f_{Sample}. These alias bands set the requirements for the sampling rate of the ADC. As can be derived from Eq. (2.9) and as depicted in Fig. 2.13, if the sampling frequency is smaller than twice the signal bandwidth, the alias bands overlap with the original spectrum, which causes information loss, as it is not possible to distinguish the original signal from the alias signal. Note that if out-of-band frequencies are present in the input signal (e.g., chopper artifacts, RF interference tones, etc.), sampling causes aliasing of these frequencies in the signal band. This explains why an AAF is required to filter out such tones (see the previous section). To avoid aliasing, the

[16] In a real ADC, the frequency spectrum of the ADC can be obtained by applying the fast Fourier transform (FFT) on the obtained sampled time-domain signal $V_{\text{s}}(t)$.

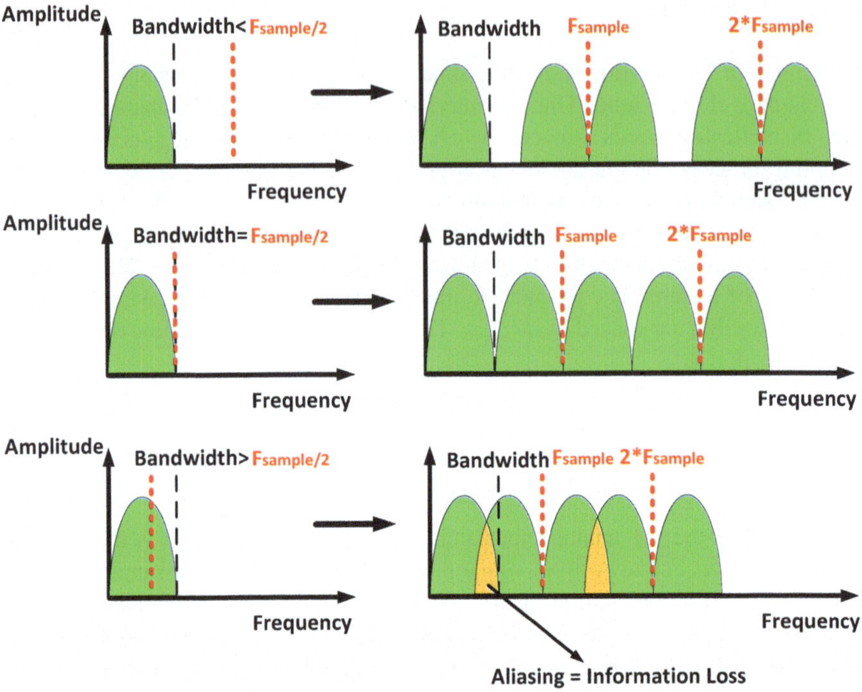

Fig. 2.13 Sampling process of a CT analog signal in the frequency domain for three scenarios ($f_{\text{Sample}} > 2*\text{BW}$, $f_{\text{Sample}} = 2*\text{BW}$, and $f_{\text{Sample}} < 2*\text{BW}$)

ADC sampling frequency should thus fulfill the following requirement:

$$f_{\text{Sample}} \geq 2 \cdot BW, \tag{2.10}$$

where BW is the signal bandwidth. This criterion is known as the Nyquist-Shannon sampling theorem:
A band-limited continuous-time signal can be sampled and perfectly reconstructed if the sampling rate is equal to or more than twice the signal's instantaneous bandwidth.

This theorem forms the basis for any sampled system, even beyond ADC circuits, e.g., switched-capacitor circuits.[17]

[17] Good to note is that, to achieve perfect reconstruction of the original signal, in a real system, the sampling frequency should be taken slightly higher than what the theorem predicts, as filters (both analog and digital) cannot implement a brick-wall type of filter, that removes all spectral content above the BW. Taking a sampling frequency above the theoretical Nyquist frequency allows for a more relaxed filter requirement.

2.4.1.2 Quantization

Quantization is the process that maps an analog amplitude to a digital code that corresponds to the nearest DT amplitude. The DT amplitude corresponding with this digital code can be written as

$$V_{\mathrm{q}} = \mathrm{LSB} \cdot \sum_{n=0}^{N-1} b_n \cdot 2^n, \tag{2.11}$$

where N is the resolution of the ADC and b_0 to b_{N-1} are the bits in the digital code. The least significant bit (LSB) is defined as

$$\mathrm{LSB} = \frac{V_{\mathrm{ref}}}{2^N} \tag{2.12}$$

with V_{ref} the ADC reference voltage, i.e., the voltage reference with which the input signal is compared. Each different possible value of V_{q} is called a quantization level.

With quantization, inevitably, a rounding error is made, which is related to the resolution of the ADC. This rounding error, which is the difference between the original input and V_{q}, is called the quantization error $\mathrm{e_q}$. For a low-frequency ramp input, this error is shown in Fig. 2.14. Quantization errors are unavoidable in digital systems and, at first sight, are deterministic in nature (e.g., for a known input, the output of an ideal quantizer will always be the same) [105]. A deterministic, nonlinear error in the time domain results in harmonic distortion in the frequency domain. However, if a sufficiently high ADC resolution (≥ 7 bits) is chosen, which makes that the input signal passes through many quantization levels, and if the ADC input frequency and sampling frequency are not related, $\mathrm{e_q}$ can be approximated as a stochastic quantity [105]. This book (and most works on ADCs) adopts this second approach; however, note that it is an approximation!

To derive the statistical properties of the quantization error, it is assumed that over a long time span every possible value of $\mathrm{e_q}$ is equally likely to occur, such that the quantization error has a uniformly distributed probability density function (PDF) $\mathrm{f_q(e_q)}$ (see Fig. 2.14). To quantify the error power of the quantization error, the variance of $\mathrm{e_q}$ can be calculated as

$$\sigma_{\mathrm{e_q}}^2 = \int_{-\mathrm{LSB}/2}^{\mathrm{LSB}/2} \mathrm{f_q(e_q)} \cdot \mathrm{e_q}^2 \cdot \mathrm{de_q} = \frac{\mathrm{LSB}^2}{12}. \tag{2.13}$$

The square root of this equation $\sigma_{\mathrm{e_q}}$ is referred to as the quantization "noise." This noise is uniformly distributed over the full Nyquist bandwidth and hence can be approximated as a white noise source. With Eq. (2.13) for the quantization noise, the signal-to-quantization-noise ratio (SQNR) can be calculated. Assuming that the ADC has a full-scale sinusoidal input with peak-to-peak amplitude V_{FS}, the SQNR

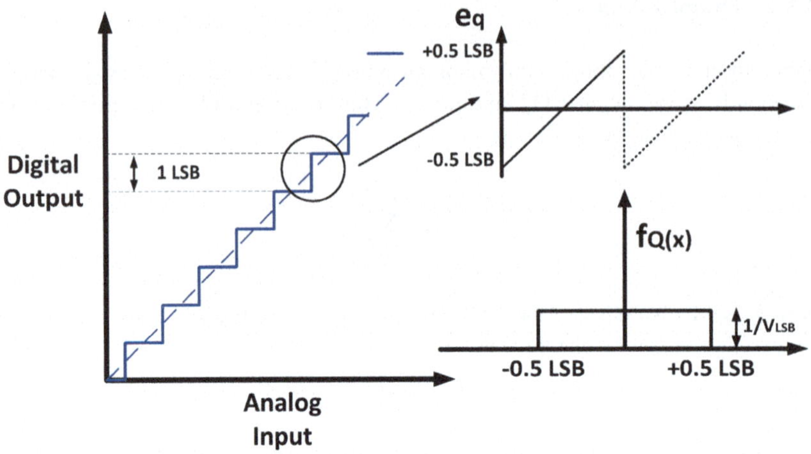

Fig. 2.14 Quantization error with uniformly distributed PDF

is equal to

$$\text{SQNR} = 10\cdot\log(\frac{P_{\text{Signal}}}{\sigma^2_{\text{e}_\text{q}}}) = 10\cdot\log(\frac{\frac{V_{\text{FS}}^2}{8}}{\frac{\text{LSB}^2}{12}}) = 10\cdot\log(1.5\cdot 2^{2N}) = 6.02\cdot N + 1.76. \tag{2.14}$$

This expression for SQNR provides a lower bound on the accuracy that an ADC can achieve.

Since the quantization noise is distributed across the full sample bandwidth, oversampling the signal with a certain oversampling ratio (OSR) can lower the quantization noise in the signal band, as can be seen in Fig. 2.15. The in-band SQNR that is obtainable via oversampling is equal to

$$\text{SQNR}_{\text{in-band,OS}} = 6.02\cdot N + 1.76 + 10\cdot\log(\frac{f_{\text{Sample}}}{f_{\text{Nyquist}}}) = 6.02\cdot N + 1.76 + 10\cdot\log(\text{OSR}) \tag{2.15}$$

with f_{Nyquist} the Nyquist frequency (i.e., twice the bandwidth).[18]

By means of oversampling in combination with noise shaping, the quantization error in-band can be reduced further, by pushing a majority of the quantization noise out of band, as can be seen in Fig. 2.15. By means of internal loop filters, the ADC achieves a noise-shaping profile. While the mathematics behind noise

[18] Note that this improvement in SQNR is achievable within the signal band of the original signal, and hence, a digital low-pass filter should be applied to the ADC output to achieve this accuracy. Moreover, to reduce the ADC digital output rate, a decimation filter typically decimates the filtered oversampling ADC output to f_{Nyquist} again [105].

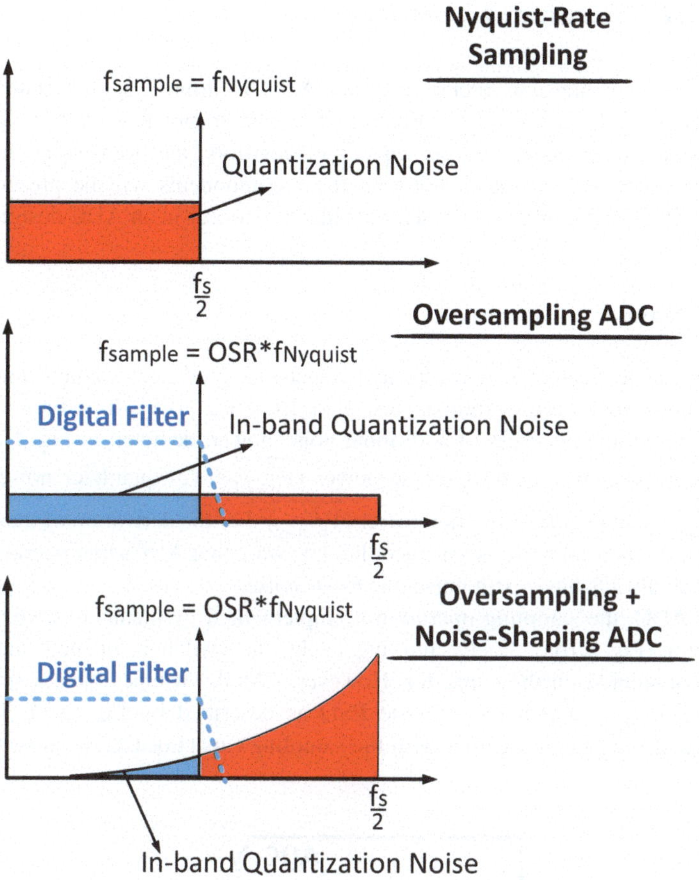

Fig. 2.15 Spectra of Nyquist-rate sampling ADC, oversampling ADC, and oversampling ADC with noise shaping

shaping would take this discussion too far,[19] it suffices to know that there are different orders of noise-shaping ADC architectures that, with increasing order, can achieve a steeper slope of noise shaping. For example, a first-order noise-shaping ADC has a quantization noise floor that increases with 20 dB/decade, a second-order architecture's quantization noise floor increases with 40 dB/decade, etc.

The quantization process is thus a fundamental limit to the ADC accuracy. Reaching this limit, however, can be difficult, since, depending on the ADC specifications, ADC accuracy limitations caused by circuits nonidealities might dominate. This will be discussed next.

[19] One of the best books for understanding noise-shaping ADCs is the classic textbook by Schreier and Temes [128].

2.4.2 ADC Accuracy Limitations

The maximum achievable accuracy of an ADC is limited by the fundamental mathematics of the quantization process. However, in practice an ADC consists of electronic components such as capacitors, resistors, and transistors. As such, additional noise and mismatch between these components will be present. This section will discuss common nonidealities that may occur in an ADC design.

2.4.2.1 Noise

As discussed in Sect. 2.3, resistors and transistors produce a certain amount of thermal noise and, at low frequencies, also 1/f noise [105]. In a typical ADC, there are three main sources of additional noise power, depicted in Fig. 2.16b: (1) sampling noise power or kT/C-noise power ($\overline{V^2_{\text{n,kT/C}}}$), (2) quantizer noise power ($\overline{V^2_{\text{n,quantizer}}}$), and (3) aperture jitter power ($\overline{V^2_{\text{n,jitter}}}$). For the derivation of these noise sources, only thermal noise is assumed, but be aware that ADCs that operate at low frequencies always have extra noise due to 1/f noise.

In an ADC, the sampling operation is implemented by means of a switch and a capacitor (Fig. 2.16a). Each sampling cycle, the switch is opened and stores the current sample on the capacitor. However, CMOS switches have a finite ON resistance R_{switch}, which has a noise PSD as described by Eq. (2.1). The ON resistance of the switch together with the sampling capacitor C_{Sample} forms a low-

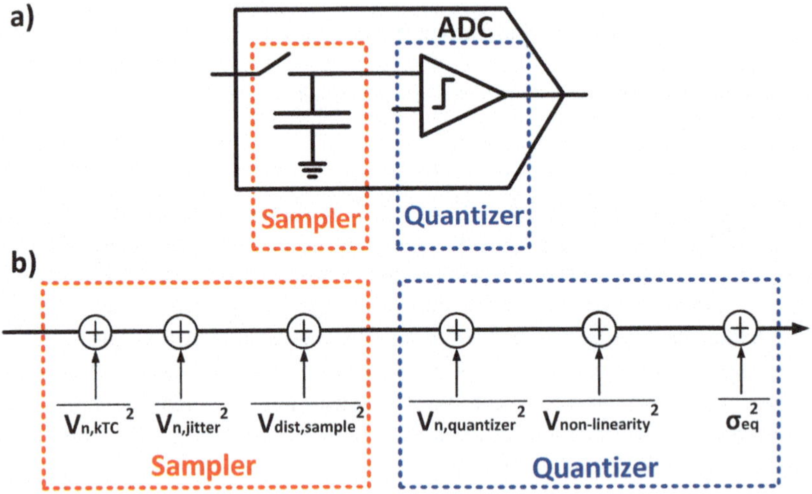

Fig. 2.16 (**a**) Block diagram of the two main ADC functional blocks: the sampler and the quantizer and (**b**) simplified model of the different ADC error sources

pass filter with a cut-off frequency equal to

$$f_{\text{cut-off}} = \frac{1}{2\pi \cdot R_{\text{switch}} \cdot C_{\text{Sample}}}. \tag{2.16}$$

The noise from the switch is filtered out by this filter, which results in the following expression for the total integrated sampling noise power:

$$\overline{V^2_{\text{n,kT/C}}} = v_n^2 \cdot \text{NBW} = v_n^2 \cdot f_{\text{cut-off}} \cdot \frac{\pi}{2} = \frac{k \cdot T}{C_{\text{Sample}}}, \tag{2.17}$$

where NBW is the equivalent noise bandwidth of the filter ($= f_{\text{cut-off}} \cdot \frac{\pi}{2}$), which is used to model a "brick-wall" characteristic of the first-order filter. From Eq. (2.17), it becomes clear that the integrated sampling noise power depends only on the value of the sampling capacitor. For medium-resolution (<10 bits) ADCs, the sampling noise power is typically taken as a fraction (e.g., 10%–30%) of the quantization noise power. The sampling capacitor C_{Sample} acts as a capacitive load on the ADC buffer. Hence, the choice of the sampling capacitor also influences the drive requirements of the ADC buffer. Therefore, increasing the ADC resolution increases not only the ADC power but also the power of the ADC buffer.

A next noise source is the quantizer noise. Assuming that the quantizer is limited by thermal noise, the quantizer noise power can be derived by modeling the quantizer as an integrator, which integrates the noise on a load capacitor C_{load}. This leads to (approximately) the following expression for the quantizer noise power [10]:

$$\overline{V^2_{\text{n,quantizer}}} = \frac{2 \cdot k \cdot T \cdot n}{A_v \cdot C_{\text{load}}} \tag{2.18}$$

with A_v the gain of the quantizer and n the weak-inversion slope factor, which is a technology constant. This expression shows again how the thermal noise depends on a capacitor that acts as a load. As the quantizer has to continuously charge and discharge this capacitor, a lower noise requirement leads again to a higher power consumption of the ADC.[20] Like the sampling noise, the quantizer noise is chosen as a fraction of the quantization noise power for medium-resolution ADC designs.

For completeness, we also discuss the aperture jitter noise, which is mostly of importance in high-speed ADCs and less in biomedical systems.[21] The sampling operation of the ADC switch is triggered by a reference signal: the clock signal.

[20] While thermal noise sets the minimum requirements for the quantizer power, if 1/f noise dominates the noise budget, devices might have to be upsized, which would lead to a larger C_{load} and thus excess power consumption.

[21] This is only true to a certain extent, as an extremely noisy clock source eventually introduces jitter errors. But, in practice, if a clean reference source like, e.g., from a crystal oscillator is used, the jitter of the on-chip clock distribution likely does not dominate the noise contributions.

Such a reference signal is often generated by (an off-chip) reference like a crystal oscillator and then converted to the proper sampling frequency by an on-chip phase-locked loop. From the output of the phase-locked loop, a clock distribution network routes the clock signal to the ADC [118]. While crystal oscillators can provide a very precise, low-noise timing reference, the phase-locked loop and clock distribution network add "clock jitter," which is an uncertainty on the arrival of the clock edges. This uncertainty on the clock signal translates to an equivalent noise voltage that is sampled each time the ADC takes a sample [118]. The larger the slope of the input signal, the larger this voltage error. For a sinusoidal input with peak-to-peak amplitude of V_{FS}, the integrated jitter-induced noise power becomes [105]:

$$\overline{V_{\text{n,jitter}}^2} = \frac{1}{2} \cdot (\pi \cdot f_{\text{in}} \cdot V_{\text{FS}})^2 \cdot t_{\text{jit}}^2, \tag{2.19}$$

where t_{jit} is the total root-mean-square (RMS) timing error due to jitter. In the best ADC designs, this error can be reduced to below 100 fs (which suffices for biomedical systems). However, further reducing it is a serious challenge in high-speed designs [93].

2.4.2.2 Systematic Errors

Noise sources add uncertainty to each ADC measurement. However, besides noise, which is a statistical fluctuation, there can also be more systematic errors in the conversion of the ADC. These are errors that occur even when averaging out many ADC samples for the same input value. These errors make that the ADC output differs from the ideal value given by the transfer function (TF) in Fig. 2.14. In Fig. 2.17 the effect of the various systematic errors on the ADC TF is shown.

A first type of error is the offset error. These errors cause a DC shift (positive or negative) in the ADC TF. In a practical ADC, this offset can be found by applying a slow ramp, from DC up to the point at which the ADC first changes an LSB. Once the offset error is known, it can easily be corrected by subtracting it digitally from the ADC output code. Offset errors can be linked to the offsets of the ADC components, such as, for example, the quantizer offset. Note that while the offset error can be corrected in the digital domain, if it is not compensated in the analog domain, it can reduce the DR of the ADC [105].

A second error that commonly occurs is a gain error. The ideal ADC normally implements a perfect linear TF with a gain of 1. In a real ADC, where a quantizer compares an input signal with a reference, parasitics in the ADC may reduce the full-scale voltage of the ADC, which causes the gain to deviate. If a negative gain error occurs (thus a gain <1), the ADC output is compressed (i.e., the ADC output does not use the full scale). If a positive gain error occurs, the ADC's input range is reduced. A simple first-order correction for the gain error can be found by finding the input for which the ADC reaches the full scale (in case of positive gain error) or applying the full-scale input and seeing which ADC output value corresponds to

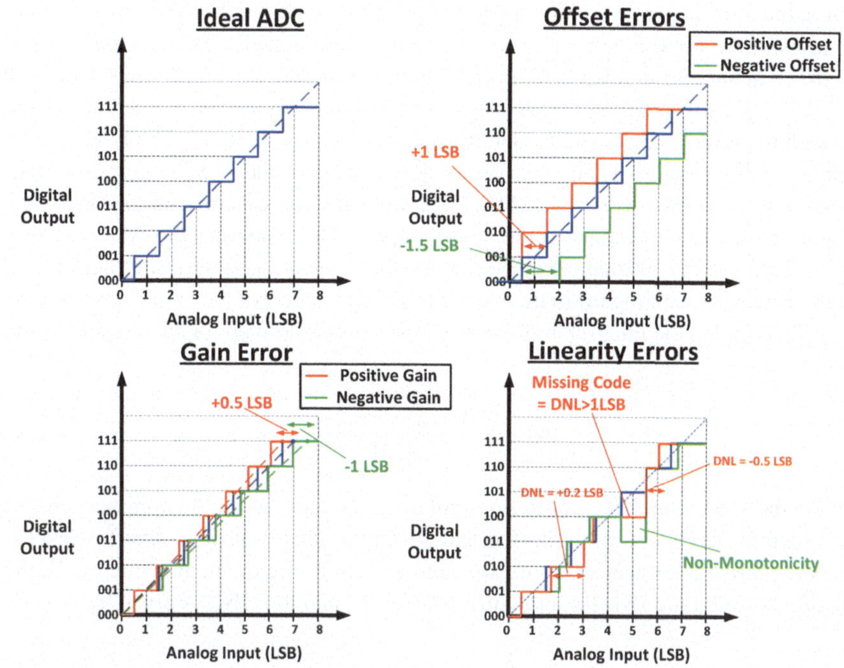

Fig. 2.17 ADC TF and common ADC errors: ideal TF (left above), ADC TF with offset error (right above), ADC TF with gain error (left below), and ADC TF with nonlinearities, missing codes and non-monotonicity (right below)

this. With both the offset error and the full-scale value known, the linear gain of the ADC can be derived and corrected for by digitally multiplying the ADC output with the inverse of the gain.[22]

A third type of errors are the nonlinearity errors (see also $\overline{V^2_{\text{non-linearity}}}$ in Fig. 2.16b). After the offset and the gain error of an ADC have been corrected, there still remain errors in the quantization levels of the ADC. Ideally, each quantization level is spaced 1 LSB apart, but, in practice, each ADC quantization level can deviate from this. The deviation between two neighboring quantization levels is referred to as the differential nonlinearity (DNL). The DNL of quantization level $i + 1$ (with$i = 0, ..., N - 2$) is described as the step width between the two subsequent quantization levels $i + 1$ and i:

$$\text{DNL}_\text{i} = \frac{V_{\text{i+1}} - V_\text{i} - \text{LSB}}{\text{LSB}}. \tag{2.20}$$

[22] Note that in Fig. 2.17 the gain error is depicted as a shift in the linear transfer characteristic, but it might occur as well that the TF becomes nonlinear and that square or cubic terms are introduced.

Often, the most negative and the most positive DNL value of the ADC are reported, as an estimate for the worst-case nonlinearity. Note that if a DNL value of -1 occurs, a quantization level is skipped, which is referred to as a missing code (see Fig. 2.17). Ideally, a circuit designer tries to avoid missing codes, as such errors are difficult to correct and, hence, lead to degradation in ADC accuracy. If a DNL occurs that is smaller than -1, an even worse degradation of the ADC occurs: the non-monotonicity. If this occurs, the ADC's TF does not increase in a monotonic way, which maps different inputs to the same output code, rendering the ADC useless. An alternative way of quantifying the nonlinearity errors is describing them based on the deviation with regard to the ideal TF. This deviation of the "real" TF from the ideal TF is called the integral nonlinearity (INL) and is defined for each quantization level i as

$$\mathrm{INL_i} = \frac{V_{\mathrm{i,real}} - V_{\mathrm{i,ideal}}}{\mathrm{LSB}}. \tag{2.21}$$

Unlike the DNL, this INL metric is cumulative in nature, as all the nonlinearities of the previous levels contribute to the shift from the ideal transfer curve for level i. As consecutive nonlinearity errors are summed up (hence the name "integral"), the relation between the INL for a quantization level j and the DNL is given by

$$\mathrm{INL_j} = \sum_{i=0}^{j-1} \mathrm{DNL_i}. \tag{2.22}$$

In a practical ADC, it is easier to calculate the DNL, as this is related to the difference between measurable quantization levels; the INL can then be calculated with the above formula. As the INL provides a metric on how the ADC deviates from the ideal TF, experienced circuit designers can use it to determine the main error source of the ADC. For example, in a simple binary weighted DAC, the largest INL error occurs near the midpoint, as this point involves a jump in the DAC code with the largest accumulated error within such design. Like the DNL, the worst-case INL is often reported in the datasheet of an ADC.

The INL and DNL are metrics to describe the nonlinearity of the ADC. It is, however, important to note that they are DC metrics, e.g., for a DC or low-frequency input of the ADC. In an ADC that converts higher frequency AC signals (i.e., almost all ADCs in practice), the nonlinearity of the ADC causes harmonic distortion. In the following section, the relation between the harmonic distortion and the INL is described in more detail.

2.4.2.3 Sampling Distortion

The last type of errors that can occur in an ADC is sampling distortion (see $\overline{V^2_{\mathrm{dist,sample}}}$ in Fig. 2.16). Different from systematic errors, sampling distortion is related to artifacts occurring in the sampling operation and can be frequency dependent. This distortion originates from various characteristics of the CMOS switches that serve as sampling switch for the ADC. The discussion below is limited to a few distortion sources: charge injection and kickback, variable switch resistance, and hold-phase droop.

Each time a sample is taken by the ADC, the sampling switch opens. As the switch is implemented by transistors, charge flows in the channel. After the signal is sampled, the switch closes, but the charge in the channel Q_{channel} does not instantaneously disappear and needs to flow either back into the signal source or onto the sampling capacitor. This introduces a certain voltage drop on the sampling capacitor (also called pedestal error), which leads to distortion if not constant [105]. To avoid charge injection, bottom-plate sampling can be used [105]. A similar effect occurs when the switch is turned ON. Due to a large voltage swing at the gate of the switch, crosstalk through the parasitic gate-drain and gate-source capacitors may also cause a voltage drop or may cause a voltage drop at the input of the ADC. This is referred to as kickback.

The ON resistance of MOS switches needs to be designed such that the switch bandwidth, Eq. (2.16), is several times larger than the signal bandwidth, such that the sampled signal can settle with sufficient accuracy. However, this resistance depends on the Vgs of the switch and, hence, depends on the signal's amplitude. To overcome this issue, transmission-gate switches (NMOS and PMOS switches in parallel) can be used, but these switches still show some amplitude-dependent resistance. Therefore, most modern ADCs use a bootstrap switch [116], which provides a constant Vgs for the switch by increasing the gate voltage proportionally to the input voltage.[23]

A last source of sampling distortion is the hold-phase droop. The charge on the sampling capacitor leaks away during the ADC conversion, due to leakage in the switch, gate leakage in the quantizer, etc. The slower the conversion, the more outspoken this effect becomes, and the larger the distortion in the ADC. However, a slow ADC conversion would allow for the sampling switch channel length to be larger than the minimum size, which can decrease the switch's leakage.[24]

[23] Bootstrapping works well at low frequencies, but at high input frequencies, a small increase in switch ON resistance is noted, as the bootstrap feedback loop cannot track the signal sufficiently fast anymore [113, 114].

[24] While droop in most conventional, clocked ADCs might not be the most important distortion source, we will see in the next chapters that for level-crossing ADCs, leakage becomes a much more important source of error, at very low input frequencies. Level-crossing ADCs have to be able to work for a small period of time at a high sample frequency. Hence, the device sizes should be small. However, most of the time, the level-crossing ADC will have a low-frequency input and work in an idle mode. Over time, charge leaks away, leading to distortion for very slow input signals.

2.4.3 ADC Accuracy Metrics

In the previous subsection several accuracy limitations of an ADC such as noise and nonlinearities were discussed. While nonlinearities can be described by the earlier introduced INL/DNL metrics, which are time-domain DC parameters, the ADC performance is typically described in the frequency domain and with AC signals. This subsection discusses several metrics that are typically used for ADCs.

2.4.3.1 Harmonic Distortion

An ADC exhibits deterministic errors, which translate into a nonlinear transfer curve in the time domain. This nonlinear transfer curve in the time domain introduces harmonic distortion in the frequency domain. For a single-tone input, Fig. 2.18 shows the typical ADC spectrum. Due to the nonlinearity of the ADC, the input tone (the fundamental tone) generates harmonic tones at multiples of f_{input}. The ratio of the input tone to these harmonics is referred to as HD2, HD3, HD4, etc. These harmonic tones can be linked to the INL of the ADC by the following approximation:

$$\text{HD}_{\text{max}} \approx 20 \cdot \log(\text{INL}_{\text{max}}/2^N), \tag{2.23}$$

where INL_{max} is the maximum INL of the ADC and HD_{max} is the largest harmonic distortion ratio. Each distortion tone has a certain distortion tone power, P_{HDx}. In practical applications, the harmonics up to the seventh harmonic carry most of the signal power, and hence the other harmonics can be neglected.

2.4.3.2 Spurious-Free Dynamic Range

Each ADC exhibits harmonic tones at the output. To quantify the spectral purity (e.g., the AC linearity) of the ADC, an often used metric is the spurious-free dynamic range (SFDR). The SFDR is defined as the ratio between the input tone signal power and the harmonic tone with the highest power. For example, in Fig. 2.18a the third harmonic is the dominant harmonic. Mathematically the SFDR is defined as

$$\text{SFDR} = 10 \cdot \log\left(\frac{P_{\text{Signal}}}{P_{\text{HD}_{\text{max}}}}\right). \tag{2.24}$$

Often, the effectiveness of a calibration technique is expressed in the improvement in SFDR.

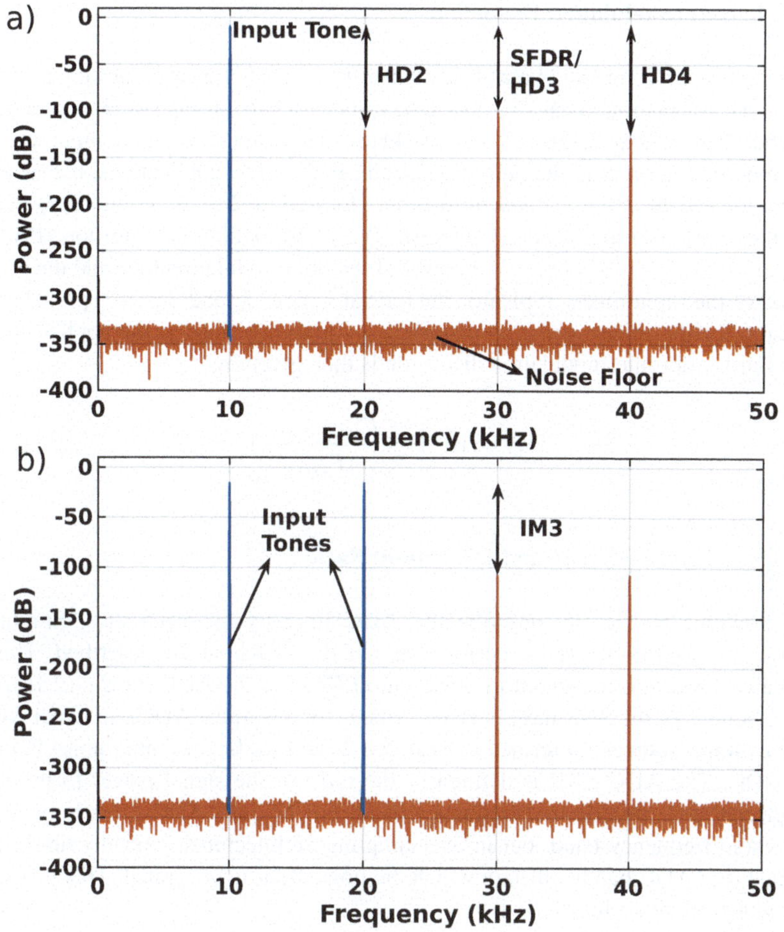

Fig. 2.18 ADC spectrum from a 10-bit, 100-kSamples/s ADC for (a) a single-tone sine input of 10 kHz and (b) a dual-tone input of 10 and 20 kHz

2.4.3.3 Total Harmonic Distortion

Related to the SFDR is the total harmonic distortion (THD). This metric is the ratio of the total power of all harmonic tones (thus, in practical applications the sum of P_{HD2} to P_{HD7}) to the power of the fundamental tone. It can be expressed both as a percentage (which is often done in audio applications) and as a logarithm. The THD, as a logarithm, is given by

$$\mathrm{THD} = 10 \cdot \log(\frac{P_{\mathrm{Total,Distortion}}}{P_{\mathrm{Signal}}}) = 10 \cdot \log(\frac{\sum_{i=2}^{\mathrm{inf}} P_{\mathrm{HD_i}}}{P_{\mathrm{Signal}}}). \tag{2.25}$$

2.4.3.4 Intermodulation Distortion

To give a more complete characterization of the ADC's linearity, not only single-tone tests are performed but also dual-tone tests with two closely spaced frequencies (in the case of Fig. 2.18b 10 and 20 kHz). This dual-tone tone input creates intermodulation tones in the output spectrum. In practical applications, the second-order (located at $f_1 \pm f_2$) and third-order (located at $2.f_1 \pm f_2$ and $2.f_2 \pm f_1$) intermodulation products are of interest. The intermodulation distortion ($\mathrm{IM_x}$) is then given as the ratio between the power of the intermodulation tone and the power of one of the input tones. Typically, the tone at $2.f_1 - f_2$ and $2.f_2 - f_1$ is taken as measure for the intermodulation distortion as these are close to the applied tones. This third-order intermodulation distortion is then given by

$$\mathrm{IM_3} = 10 \cdot \log(\frac{P_{2 \cdot f_{2/1} + f_{1/2}}}{P_{f_{1/2}}}). \tag{2.26}$$

2.4.3.5 Signal-to-Noise(-and-Distortion) Ratio

The previous metrics all describe the ADC linearity characteristics. Similarly, metrics to describe the noise performance of the ADC can be described. This is typically done with the signal-to-noise ratio (SNR) of the ADC. In Eq. (2.14), the lower bounds of the SNR have been described (for a Nyquist ADC). In a real ADC, various noise sources (described in Sect. 2.4.2) add to the minimum noise floor of the ADC. The ADC SNR is defined as the ratio of the signal power to the total integrated noise in a desired frequency band. For Nyquist ADCs, this is typically the whole frequency band, but in oversampling architectures, typically an f_{BW} is defined (the ADC bandwidth), in which the noise floor is integrated. This gives the following definition for SNR:

$$\mathrm{SNR_{ADC}} = 10 \cdot \log(\frac{P_{\mathrm{Signal}}}{P_{\mathrm{Noise}}}) = 10 \cdot \log(\frac{P_{\mathrm{Signal}}}{\int_0^{f_{\mathrm{BW}}} NSD(f) \cdot df}), \tag{2.27}$$

where $NSD(f)$ is the noise spectral density of the ADC (the noise power per unit bandwidth at a certain frequency). In some applications, the effect of both noise and nonlinearities is undesired. Hence, a metric called signal-to-noise-and-distortion ratio (SNDR) is used to describe the ADC accuracy. This metric is often used in the ADC FoM (see below), because it entails all error sources accuracy in an ADC. The SNDR is defined as

$$\begin{aligned} \mathrm{SNDR_{ADC}} &= 10 \cdot \log(\frac{P_{\mathrm{Signal}}}{P_{\mathrm{Noise}} + P_{\mathrm{Total,Distortion}}}) \\ &= 10 \cdot \log(\frac{P_{\mathrm{Signal}}}{\int_0^{f_{\mathrm{BW}}} NSD(f) \cdot df + \sum_{i=2}^{\mathrm{inf}} P_{\mathrm{HD_i}}}). \end{aligned} \tag{2.28}$$

Related to SNDR, sometimes the effective number of bits (ENOBs) is defined. The ENOB indicates how many "actual" bits of precision the ADC has (compared to the precision predicted by quantization only). The ENOB of the ADC is given by

$$\text{ENOB} = \frac{\text{SNDR}_{\text{ADC}} - 1.76}{6.02}. \tag{2.29}$$

This definition is derived from the equation for the signal-to-quantization-noise ratio (SQNR), where the SQNR is replaced by the SNDR.

2.4.3.6 Dynamic Range

A last important metric for an ADC is the dynamic range (DR_{ADC}). This DR_{ADC} is the ratio between the largest (full-scale) input and the smallest input signal that the ADC can resolve. The DR can thus be written as

$$\text{DR}_{\text{ADC}} = 10 \cdot \log\left(\frac{P_{\text{Signal,FS}}}{P_{\text{Signal,min}}}\right) \tag{2.30}$$

with $P_{\text{Signal,FS}}$ the ADC input power for a full-scale input and $P_{\text{Signal,min}}$ the signal power of the smallest signal that the ADC can resolve.

2.4.4 ADC Topologies

This section will discuss the most popular ADC topologies for sensing applications.[25] The different ADC topologies that will be discussed are shown in Fig. 2.19.

2.4.4.1 The Flash ADC

The flash ADC (Fig. 2.19a) consists of a thermometer-code reference DAC (a resistor ladder), a sampling capacitor, a bank of comparators that compares the input sample with each reference level of the DAC, and a thermometer-to-binary encoder. The flash ADC takes a sample each clock cycle, which is then compared to all reference levels in the same cycle. The value of the sample is directly available at the comparator output, which is then converted by the encoder to a binary word. As the flash ADC finishes a comparison in a single clock cycle, it is in absolute speed terms the fastest ADC topology and hence used for high-speed designs [105].

[25] While nowadays many different ADC topologies exist and this list is far from exhaustive, the section is only intended to give the reader an appreciation on how an ADC is actually built up as a circuit.

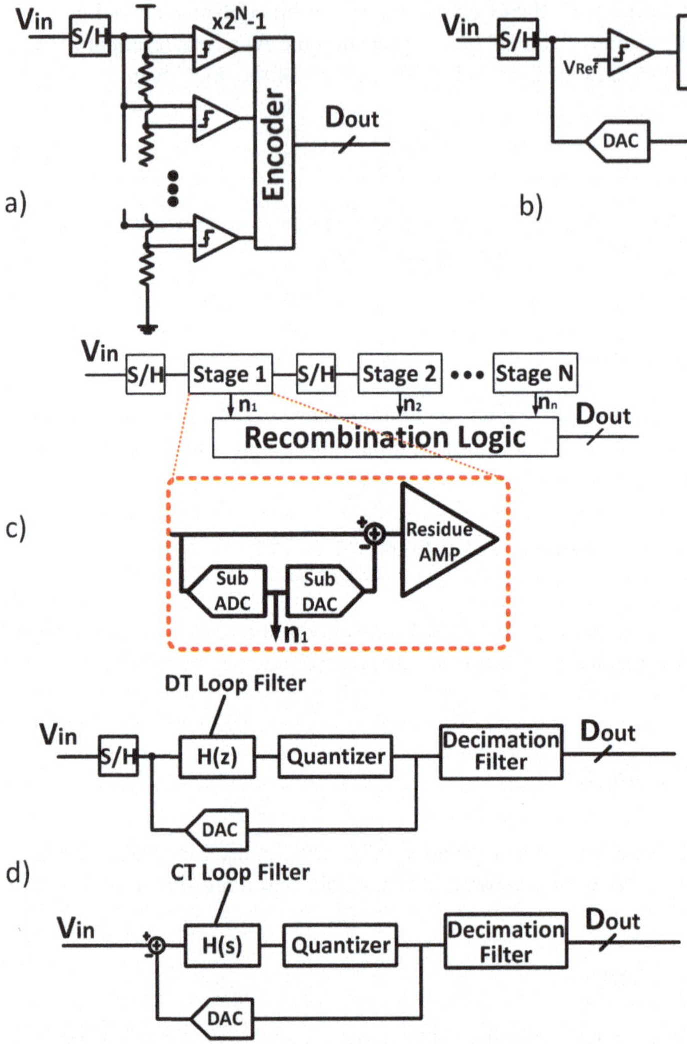

Fig. 2.19 Overview of classical ADC topologies: (**a**) the flash ADC, (**b**) the SAR ADC, (**c**) the pipeline ADC, and (**d**) the (DT and CT) delta-sigma ADC

The topology is, however, not suited for high ADC resolutions ($>$ 5 bits). As the amount of comparators grows exponentially with the resolution ($2^N - 1$ comparators are required), the power consumption of the ADC becomes prohibitively large for high resolutions [105]. Moreover, as each comparator makes a decision at the same time, the offset of each comparator should be very low, to avoid non-monotonicities in the ADC TF. This makes that power-hungry calibration for the comparators is required [105]. Moreover, if all comparators make a simultaneous decision, this causes a large kickback to the input and the reference voltage ladder, which may distort the proper ADC operation or increase the ADC buffer driving requirements (which entails a power overhead). Moreover, the matching of the resistor reference ladder becomes more stringent with higher resolutions.

2.4.4.2 The Successive-Approximation ADC

The successive-approximation ADC (SAR ADC) is given in Fig. 2.19b. It consists of a comparator, successive-approximation register (SAR), a DAC—which is often implemented as a capacitive DAC (CDAC)—and a sample and hold stage, which is in most implementations merged with the CDAC. The SAR ADC converts the analog input by doing a binary search through all quantization levels of the ADC. At each cycle, the DAC generates a reference voltage that is compared to the input signal. If the input is larger, the comparator decision is positive, and that bit is taken equal to 1. Alternatively, if the decision is negative, the bit is equal to 0. Depending on the decision of the comparator, the SAR adjusts the search range for the next comparison. This binary search is repeated for N cycles, such that an N-bit accuracy is reached. An N-bit SAR ADC, hence, converts the input in N clock cycles + the sampling cycle (which can take up more than one clock cycle) [45, 93, 105]. Historically, the SAR ADC was therefore seen as a medium-resolution, medium-speed ADC topology. However, the SAR ADC has emerged in recent years as a versatile ADC topology, which is both used in low-power, medium-resolution systems (such as biomedical systems) [189] all the way up to high-speed designs [113, 114] and even in high-resolution designs [63]. The SAR ADC benefits well from scaled CMOS technologies, as most of the building blocks of the SAR are digital in nature. Due to the scalable nature of this ADC, it can also achieve a small area footprint as well, making it also a popular sub-ADC for other topologies [45, 113]. The classical SAR ADC does require two clocks: a slow sampling clock and a fast conversion clock, which increase the complexity of the clocking circuitry. However, this can be resolved by using asynchronous SAR logic, where the conversion occurs in an asynchronous fashion, removing the need for the fast conversion clock [45].

2.4.4.3 The Pipeline ADC

The pipeline ADC (Fig. 2.19c) is an ADC topology that is used for high-speed and high-resolution applications [63, 113]. While the flash ADC offers a high speed, it is limited in resolution. The SAR ADC requires N cycles to finish its conversion. Thus, it is difficult to obtain an ADC in a single stage that has both a high speed and a high resolution. Therefore, some ADCs split the ADC conversions over multiple stages, similar to digital pipelining [45, 105]. In digital pipelining, operations are split into several subsequent stages that each perform a small computation, such that one big calculation can be executed over several cycles. The pipeline ADC applies this concept to the ADC conversion. It consists of several stages that each include a sample and hold circuit, a low-resolution sub-ADC and sub-DAC, and a residue amplifier [45, 105, 113]. In the first stage, the first bits of the ADC are converted by the first sub-ADC. Next, this low-resolution digital code is converted back to analog by the sub-DAC and then subtracted from the analog input, which results in a residue voltage. This voltage is passed to a residue amplifier, and the amplified residue is passed to the next stage of the pipeline. The same process is repeated in the second stage, while in the first stage a new sample is taken and processed. This is repeated in the different stages until the original sample reaches the last stage, after which a single conversion is finished. As each stage processes a new sample in only one or few clock cycles, this ADC can achieve a fast sampling rate [45, 105]. Moreover, by using multiple stages, the pipeline ADC can simultaneously achieve a high resolution [45, 105, 113]. However, while a fast sampling rate is obtained for such ADC, the latency for a single sample can be over N clock cycles. The pipeline ADC thus has a fast sampling rate but a higher latency compared to a single-stage ADC. The sub-ADCs that are used for the pipeline ADC are often a flash or SAR ADC [45, 105, 113]. Note that while the sub-ADCs and sub-DACs are of low resolution, their precision should still be equal to that of the total ADC precision, as errors propagate as well through the different pipeline stages. Especially the residue amplifier design should be taken care of when designing such an ADC, as this block typically consumes most power in the ADC [45, 105, 113].

2.4.4.4 The Delta-Sigma ADC

The previous ADC topologies are all categorized as "Nyquist" ADCs, as these topologies sample close to the Nyquist rate. The delta-sigma ($\Delta\Sigma$) ADC, on the other hand, uses oversampling and noise shaping to achieve very high resolutions. It is a popular topology for low-bandwidth applications that require a high resolution (e.g., sensor applications) [128]. Like the SAR ADC, the $\Delta\Sigma$ ADC has seen many different variants (incremental, MASH, SMASH, etc.) in recent years, and between these different topologies there are many different circuit implementations [128]. We limit this discussion to the two most generic topologies: the CT and DT $\Delta\Sigma$ ADC (Fig. 2.19d). Both topologies consist of a loop filter, quantizer, and feedback DAC. The quantizer, a single-bit or low-resolution ADC, converts the output of the loop filter into a digital code, which is converted to analog by the DAC and subtracted from the next sample, and the residue is passed through the loop filter. If the right filter topology is chosen, the quantization noise undergoes a high-pass characteristic, while the signal experiences a low-pass characteristic. This results in a noise-shaping spectrum [45, 128]. A digital decimation filter (often implemented off chip in the literature) then filters out the noise at higher frequencies and reduces the oversampled ADC output data rate to the Nyquist rate [128]. The main difference between DT and CT $\Delta\Sigma$ ADCs is that in DT $\Delta\Sigma$ ADCs each sample is sampled at the input of the ADC on a sample and hold capacitor, and the loop filters are implemented in DT (i.e., they are switched-capacitor filters) [45, 128]. CT $\Delta\Sigma$s do not have a sample and hold stage, and the loop filter and feedback DAC operate in the CT domain. The loop filter can be implemented as a Gm-C filter, active RC filter, etc. [105]. Depending on the application at hand, both topologies offer other advantages. The CT $\Delta\Sigma$ offers the advantage that it has inherently no aliasing, and therefore requires no AAF, as the loop filter of the ADC already filters the signal before the sampling. Moreover, it also offers an easy to drive resistive input to the ADC buffer, and the opamps in the active filters consume also less power (due to the relaxed settling requirements) compared to the opamps used in the DT $\Delta\Sigma$ ADC [45, 128]. The downside of the CT topology is that, since the feedback DAC is implemented in CT, the CT topology is more prone to timing errors, delays in blocks, transient effects, etc. The DT $\Delta\Sigma$, on the other hand, suffers less from these transient effects and timing errors, which is why they can reach higher accuracies [45, 128]. This comes at a cost of more stringent AAF requirements (which are still relaxed with respect to a Nyquist-rate ADC), more difficult to drive capacitive input for the ADC buffer and a higher opamp power consumption due to the more stringent settling requirements of the switched-capacitor circuits.

Note on Hybrid ADCs: Which ADC to Choose?
Historically, each ADC had its distinct properties that were suited well for certain applications. However, in recent years state-of-the-art ADCs have become more and more hybrid topologies [45]. Such hybrid ADC topologies combine multiple favorable properties of the different ADC topologies into a single ADC architecture. Examples include the pipeline SAR ADC, in which several low-resolution SAR ADCs are pipelined, to achieve a high-speed, high-resolution ADC topology with a compact circuit area. Another example is the noise-shaping SAR ADCs, which combine the SAR ADC with principles of DT $\Delta\Sigma$ ADCs, such that a high resolution can be achieved, with a scalable circuit architecture. Remarkably, $\Delta\Sigma$ ADCs with SAR quantizers are also employed nowadays for similar reasons. With all of these hybrid architectures, it becomes difficult to choose a proper ADC for a given application.

To make an appropriate ADC choice, the following selection strategy can be used:

- Choose the best basic ADC architecture for the application (SAR, Flash, $\Delta\Sigma$, ...) that fulfills most requirements of the application (speed, resolution, power, area, etc.)
- Identify which properties are not met. Which ADC does meet such requirements? Introduce the concept of this ADC into the basic ADC architecture chosen.
- Once the ADC fulfills all requirements, further hybridization is not required, as going more hybrid includes even more design complexity, which is to be avoided.

2.4.5 ADC Performance Metrics

To end this discussion on ADCs, some ADC performance metrics are discussed, which allow to assess the ADC performance based on three important design parameters: speed, accuracy, and power.[26] The first two parameters are typically set by the ADC application, while the power needs to be minimized in almost all applications, including in the wearable biomedical systems discussed in this book. To describe the ADC power efficiency, several FoMs exist that describe the relation between these three important requirements. Three FoMs are often used: the Walden FoM, the Schreier DR FoM, and the Schreier signal-to-noise-and-distortion ratio (SNDR) FoM.

The Walden FoM is defined as

$$\mathrm{FoM}_{\mathrm{Walden}} = \frac{P_{\mathrm{ADC}}}{2^{\mathrm{ENOB}} \cdot 2 \cdot f_{\mathrm{BW,ADC}}}, \tag{2.31}$$

with P_{ADC} the ADC power consumption, ENOB the effective number of bits (see Sect. 2.4.3), and $f_{\mathrm{BW,ADC}}$ the ADC bandwidth. The FoM is expressed in energy per conversion step (E/conv.step). This Walden FoM is used for low-to-medium resolution ADC designs (<10 bits), which are typically not limited by the noise of the circuits, and in which it has been found empirically that the power approximately doubles when increasing a bit in resolution [93].

For noise-limited designs (designs 10 bits or more), thermal noise requirements limit the ADC power consumption [93].[27] To this end, the Schreier FoMs have been defined, which assume a thermal noise-limited design.

The Schreier FoM using the DR as accuracy metric is defined as

$$\mathrm{FoM}_{\mathrm{Schreier,DR}} = \mathrm{DR}_{\mathrm{ADC}} + 10 \cdot \log(\frac{f_{\mathrm{BW,ADC}}}{P_{\mathrm{ADC}}}). \tag{2.32}$$

The Schreier DR FoM however does not include the distortion and therefore may give a skewed image of the ADC performance, as the FoM does not take into account the ADC linearity. Increasing the ADC linearity can increase the power consumption of a design [93]. Hence, most designs use the Schreier SNDR FoM:

$$\mathrm{FoM}_{\mathrm{Schreier,SNDR}} = \mathrm{SNDR} + 10 \cdot \log(\frac{f_{\mathrm{BW,ADC}}}{P_{\mathrm{ADC}}}). \tag{2.33}$$

Both Schreier FoMs are expressed in dB.

---

[26] In a real ADC design, many more design parameters need to be taken into account, but the three parameters included in the FoM of the ADCs are the most fundamental ones and allow for a good first-order estimation of what is needed in a system.

[27] Thermal noise requires that, in order to increase a bit in resolution, a $4\times$ higher power consumption is required of analog circuits.

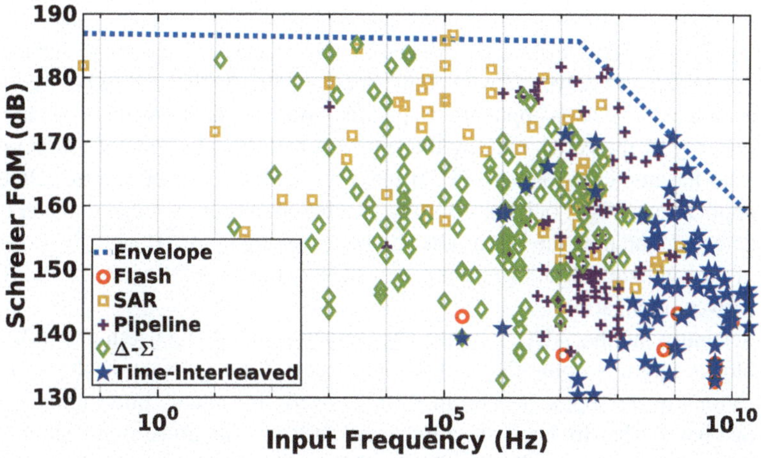

Fig. 2.20 Overview of the best published ADC topologies in the ISSCC and VLSI conference proceedings (2000–2023). Plotted is the Schreier FoM as a function of the maximum input frequency of the ADC. Note that the best FoM is 187 dB for low-frequency inputs, while for higher input frequencies, a cut-off frequency around 10 MHz is noted, after which the FoM decreases with approximately 10 dB/decade (see the plotted envelope) [92]

To keep track of the state-of-the-art power efficiency of ADCs, the Murmann chart (Fig. 2.20) [92] keeps track of the performances (and thus the FoM values) of the best ADC designs presented at the yearly ISSCC and VLSI conferences. For low-power designs, the Walden FoM can go below 1 fJ/conv.step, while the best Schreier SNDR FoM values are just below 187 dB. In Fig. 2.20, an overview (from 2010 to 2023) of the best designs of the Murmann chart is plotted. From this graph, it is clear that for the biomedical application range, ADCs can achieve very high power efficiency. Most of these ADCs are SAR ADCs [45, 92, 93].

On the (Non)Sense of ADC Figures of Merit

Historically, the power efficiency of an ADC was a bottleneck in many systems. Over the years, the circuit design community therefore focused on optimizing the power efficiency of ADCs, which led to an estimated 20 dB of improvement in power efficiency over a period of 20 years [93]. This came because of both technology scaling and architectural circuit innovations; hence, combining the two things circuit designers can rely on to improve the circuit performance [45]. However, while the effort to focus on the power efficiency clearly paid off, there should be caution when going forward with this "race to the extra decibel" in performance [93]:

(continued)

- Firstly, the ADC FoMs use SNDR as the "accuracy" metric to balance bandwidth and power. However, the accuracy metric that has to be used to evaluate the ADC is application-specific. Some applications require a high linearity of the ADC, making the spurious-free dynamic range (SFDR) or total harmonic distortion (THD) maybe a better accuracy metric. Other applications might only care for the absolute minimum voltage that can be detected, and then the DR might be a better metric to use. Hence, the FoMs of the published designs might give a skewed image of "the best" ADC for a given application.
- Secondly, as the ADC has become very power efficient, other building blocks might become the system bottleneck. Take as example the applications in this book, where an ADC converts biosignals into digital bits that have to be streamed off chip by, e.g., a Bluetooth transmitter. The best ADCs can achieve a Walden FoM of below 1 fJ/conv.step [92]. It costs thus less than 1 fJ to "generate" a bit of data, while a Bluetooth transmitter can have a power efficiency of 10 nJ/bit [187]. In such systems, the A/D conversion is almost power-free, and the greatest design effort should be on making transmitters more power efficient or on reducing the amount of ADC data (see the sections below).
- Lastly, the ADC FoM that is reported contains only the power of the core ADC circuitry and not of the auxiliary circuits that are also needed to operate the ADC, such as the input driver and the reference buffers. These circuits can easily consume as much power or more than the ADC itself [45] and, hence, should ideally also be reflected in the FoM. As certain ADCs, such as CT architectures, are easier to drive (which relaxes the input driver's power requirements), the FoM that is reported for some architectures may not lead to the lowest power in an actual integrated system.

The take-away message here is that while the FoM certainly holds value to compare different designs, at the system level, a designer should look more carefully to the actual ADC topology and the other system blocks, to determine where to put the most design effort. The ADC is not always where the most effort should be put.

2.5 Digital Signal Processor

Once the ADC has converted the biosignals into bits, the digital signal is available for the DSP. The DSP in the biomedical system is often a microcontroller unit with dedicated filters, accelerators, and processing circuits [31, 59, 158, 159]. It fulfills many tasks, from implementing the control signals for the AFE to calibration of the ADC to implementing dedicated signal processing algorithms. Hence, the microcontroller often consumes a significant amount of power in the system, certainly if older CMOS technologies are used. The latter is often the case for biomedical systems, which favor older technologies for their advantageous analog properties, their robustness, and their cheaper cost to produce [59, 158, 159]. In the Sect. 2.8 of this chapter, a more in-depth discussion of different signal processing techniques will follow. After processing the signal data, the DSP sends the data to the TX.

2.6 Transmitter

As discussed earlier in the chapter, for the comfort of the patient, a biomedical system needs to avoid using wired communication, which is why most systems use some form of (short-range) wireless communication.[28]

The system should be able to connect with a base station such as a laptop or a cellphone, where the data are either stored, or such a base station acts as a gateway to a cloud server (see Chap. 1). The power consumption of a transmitter is usually much higher compared to that of the ADC or AFE; often the TX consumes the most power of all circuit blocks [59, 158, 159, 187] (see also Chap. 1). In Fig. 2.21, a generic transmitter architecture is given, as well as some typical system-level requirement for the TX: the maximum data rate, the link range, the link robustness, and energy efficiency.

As is clear from Fig. 2.2, different signals have a different Nyquist frequency, and hence, the data rate, and thus the TX power, depends on the signal. In addition, depending on the application, the system might also require multiple recording channels, which increases the total data rate further.

Another application requirement is over which distance the wearable should be able to communicate. This is rather a practical matter, as the link range indicates how far the receiver can be placed away from the system. In some scenarios, the TX should be able to reach the receiver, even if it is stationed a bit further away. For example, if the base station is a laptop, it may not be practical to be in short range at all times. Linked to this requirement is the link robustness. Ideally, a user of a wearable should not care too much that the device has a direct line of sight to the receiver [187]; rather, the communication link should work even if objects are standing in between the receiver and the system. Moreover, with the RF spectrum becoming more and more occupied by different applications, technologies like ZigBee or Bluetooth, which use direct-sequence spread spectrum and frequency-hopping spread spectrum to decrease the interference, might be preferred [90, 129].

[28] While a detailed overview of different transmitter circuits and technologies is beyond the scope of this book, this section provides some high-level overview of some frequently used technologies, which aids a system designer in choosing the right TX circuit.

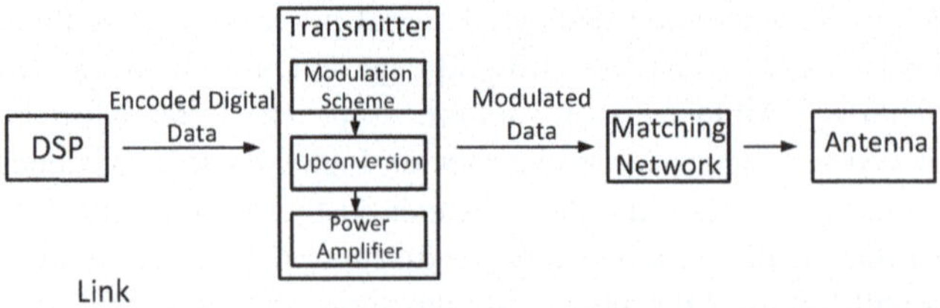

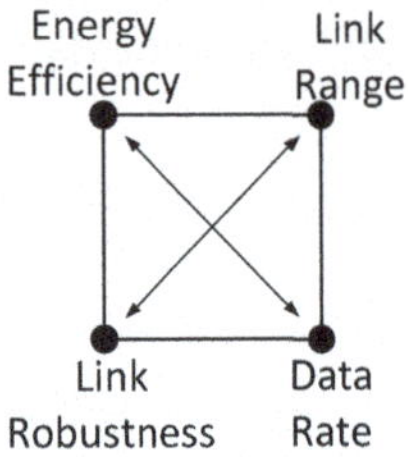

	Energy Efficiency	Link Range	Link Robustness	Data Rate
UWB	0.1-10 nJ/bit	< 10 m	++	Up to 200 Mbps
Bluetooth	1-100 nJ/bit	< 10 m	+++	Up to 2 Mbps
Zigbee	1-200 nJ/bit	< 75 m	+++	Up to 250 kbps
HBC	10pJ-1 nJ/bit	< 1.5 m	+	Up to 100 Mbps

Fig. 2.21 Overview of a generic transmitter architecture (top) and overview of the design trade-offs between various short-range communication protocols typically used for biomedical devices (bottom)

A last important parameter is the link energy efficiency, which is typically reported as the energy consumed per bit that is transmitted. As the energy budget of the wearable is set by the application, which dictates the required battery lifetime and the size of the battery, the energy efficiency together with the data rate of the application determines if a specific communication technology is suited for that application. Often, however, existing protocols alone fail to meet such requirements [59, 158, 159, 187], and since the progress in energy efficiency is improving more slowly than, e.g., that of ADCs or other analog circuits, perhaps another path that can be taken is to employ data reduction techniques in the DSP or already in the ADC, to lower the data rate.

2.7 Application Example: A 3-Lead ECG Wearable

The above sections have reviewed the different building blocks of a typical biomedical system. As became clear, to enable a long battery lifetime, the power consumption of the system should be minimized. Simultaneously, the circuits should achieve a sufficiently high recording SNR, input impedance, etc. Therefore, a small system-level case study will be presented in this section, to indicate how the different blocks of the system compare to each other in terms of power consumption.

The application at hand is the monitoring of ECG in the home environment, for patients recovering from heart surgery. The patient should continuously be monitored, and a Bluetooth TX should send the patient data to a cellphone. The application requirements are listed in Table 2.1; the values are taken from various publications that implement similar systems [43, 59, 158, 159]. The wearable should record long periods of high-quality data and, hence, has three recording channels

Table 2.1 Overview of the specifications for the 3-lead ECG example

Specification	Value
Amount of channels	3
Signal bandwidth	500 Hz
Input-referred noise LNA	1.5 μV_{rms}
NEF LNA	2
ADC resolution	10 bits
Walden FoM ADC	100 fJ/conv.step
DSP power	120 μW/MHz
Efficiency BLE TX	10 nJ/bit

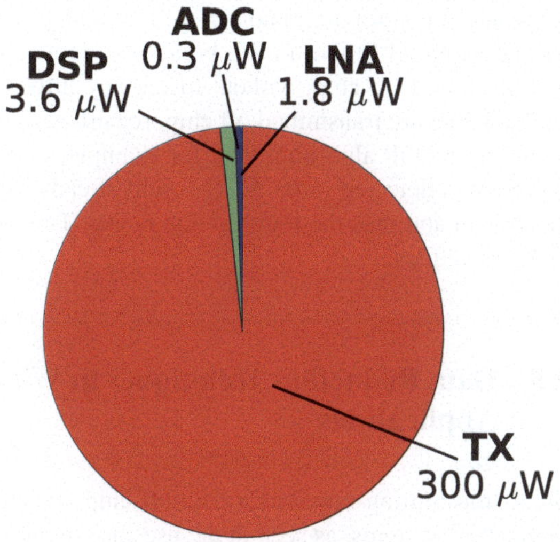

Fig. 2.22 Power breakdown of the 3-lead ECG system. Clearly, the TX consumes the most power within the system

that sample the ECG signals at a rate of 1k sample/s. To record signals with a sufficiently high accuracy, an IRN of 1.5 μV_{rms} is assumed for the LNA. Moreover, a NEF of 2 is assumed for the LNA. This is a realistic assumption for an LNA in a system, where not every building block can be optimized to the absolute smallest obtainable FoM. Similarly, the ADC is assumed to have a Walden FoM of 100 fJ/conv.step. To have sufficient DR, the ADC's resolution is chosen to be 10 bits. For the case study, it is assumed that the data are processed off chip and the on chip DSP is assumed to only serialize the data. The power of the DSP is difficult to estimate, as this highly depends on the architecture that is implemented and the functionalities that it has. Therefore, the power value for the DSP is taken from the processor reported in [159]. For the DSP operating frequency, a value of 30 kHz is taken, which would be sufficient to serialize all ADC data in this use case. Finally, for the Bluetooth TX, an efficiency of 10 nJ/bit is taken, which is a representative value taken from the literature [187].

In Fig. 2.22, the power breakdown of the system is depicted. The power of the Bluetooth TX module dominates the system power budget, and it consumes more than 150× the power of the LNA and 1000× more than the ADC. Clearly, the cost of the system is mainly in the transmission power, while at the same time, the other building blocks of the system are orders of magnitude more power efficient. To extend the battery life of the system, this TX power consumption thus needs to be tackled. One option is to improve the efficiency of the TX. However, the progress in power efficiency for such circuits is slower compared to the other analog circuits in the system [187]. Another option is to look into alternative TX technologies. In the proposed application, changing the Bluetooth TX to, e.g., a human body communication (HBC) link could yield a significant improvement. However, this would interfere with other aspects of the applications, as HBC has a smaller link range and is less robust compared to Bluetooth [43, 92, 187]. The last option, which will be explored further in this book, is to reduce the amount of data that need to be transmitted off chip, without losing any information. For example, in this use case, all data are transmitted off chip, regardless of the physical state of the patient. Including a DSP algorithm that, for example, only transmits data whenever some symptom is detected in the ECG would already drastically reduce the data rate of the system and thus the transmission power. The next section will dive deeper into such techniques.

2.8 Data Reduction Techniques in Wireless Sensing Applications

Since transmitting raw sensor data off chip consumes the most power in wireless biomedical systems, as seen in the use case in the previous section, many works in the literature have focused on reducing the data rate of such systems by including on-chip data reduction techniques. In this section, several approaches from literature will be surveyed and the advantages and disadvantages discussed.

Many natural signals, including biosignals, have a physical bandwidth, determining the Nyquist rate, which is higher than the *information rate*, i.e., the rate at which the relevant sensor data changes [152, 161]. For example, when taking images of a static scene, an image sensor generates many frames, which contain redundant data [67]. Another example is AP signals, in which the information is contained in the spikes, that only occur sporadically [183], or ECG, where the relevant information is contained in the P-QRS-T complex and not in the signal in between [152]. These signals are said to be "sparse," i.e., the signal can be represented with fewer samples (and thus compressed) than what is prescribed by the Nyquist theorem, see Eq. (2.10). By exploiting the information sparsity of the signal, the required TX bandwidth can be reduced from the Nyquist rate to the actual information rate (see Fig. 2.23), which in turn leads to significant energy savings. The trade-off that occurs when reducing the data rate to the information rate is that the system might have a loss of generality [161]. This can be understood as follows. Different classes of signals exhibit sparsity in different bases. For example, ECG

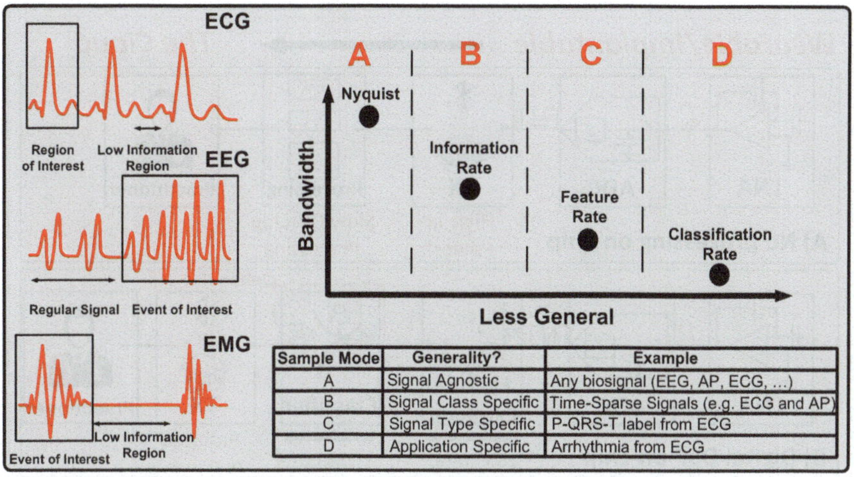

Fig. 2.23 Sparsity in sensor signals (left) and the bandwidth required by various types of signal information (right)

or AP shows burst-like behavior over time, i.e., they are compressible in the time domain. Signals like EMG, however, exhibit sparsity in the frequency domain [32]. To compress the ECG/AP or EMG signals, a different compression method is thus required.

Reducing the data rate of a system to the information rate thus helps to lower the system power. Signals that are reduced to their information rate can be fully recovered, i.e., with the compression of the signal, no signal quality or information is lost [32, 161]. This is known as *lossless* compression. As shown in Fig. 2.23, if the application allows to sacrifice even more generality, the data rate can be further reduced. This is called *lossy* compression. Often, wearable applications do not need to capture the full signal; rather, the interest is in certain signal features only [161]. Such features are a subset of the full information content of the signal, and the rate at which they change is called the *feature rate*. For example, for ECG signals, extracting the P-QRS-T complex may be sufficient to detect, e.g., heart rhythm failures [182].

From Fig. 2.23, it appears that most data can be reduced when the system reduces the data to the *classification rate*, i.e., the system only transmits the classification result of the signals. While this reduces the TX bandwidth considerably, a classification problem is always dedicated to a certain application and hence is not generally applicable.[29]

[29] Besides reducing the TX power of the system, including classification algorithms on chip also allows to reduce the closed-loop latency of a system, which is needed for, e.g., neural recording and neural stimulation applications [136], as will be discussed more in detail in Chap. 5.

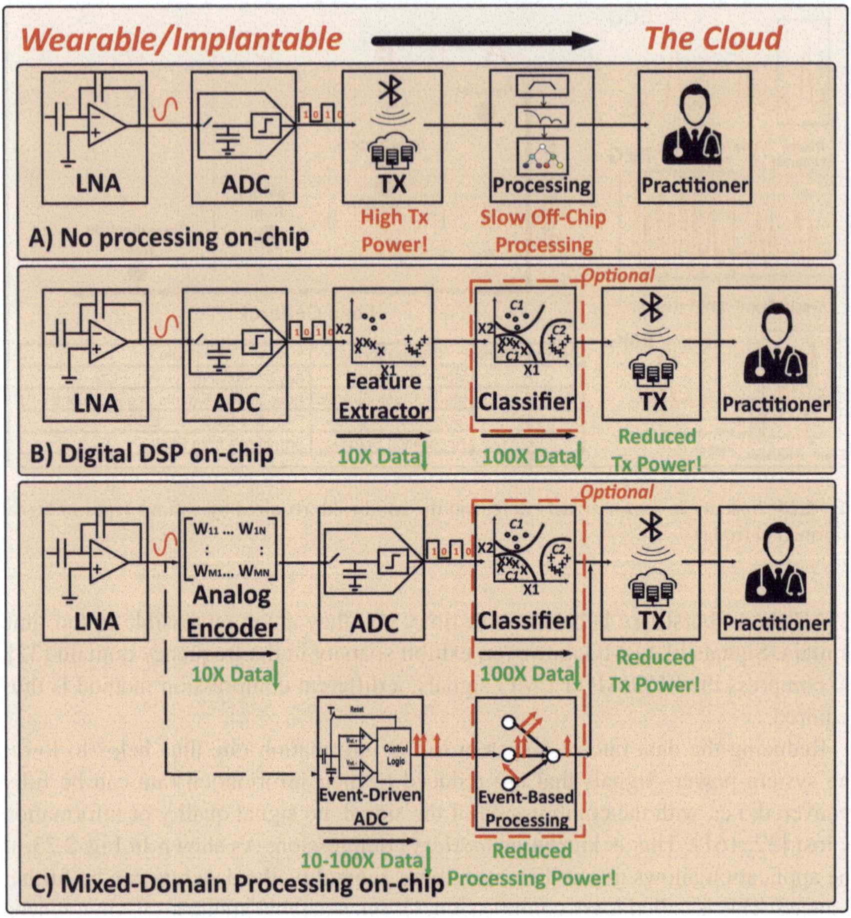

Fig. 2.24 Illustrative overview diagram of various biomedical systems with different methods of local data processing: (**a**) a system with no processing on chip, (**b**) a system with digital feature extraction and (optional) on-chip classification, and (**c**) mixed-signal data reduction techniques: compressive sensing and event-driven sampling and processing

In Fig. 2.24, several approaches to on-chip data reduction are depicted. In the next subsections, these will be discussed in more depth, and some examples from the literature will be provided to illustrate them.

2.8.1 *Systems with On-Chip Digital DSP*

A first option to reduce the sensor data is to reduce them in the digital domain (see Fig. 2.24). We consider three types of techniques: digital compression, digital feature extraction, and digital classification.

2.8.1.1 Digital Compression Algorithms

Digital compression algorithms are widely used when data transmission needs to be reduced, but the original signal needs to be able to be reconstructed at the receiver side. Compression algorithms can be implemented on an embedded microcontroller in software [76], or a dedicated hardware implementation can be used [30]. In this discussion, the focus will be on three classes of compression that are often used in biomedical systems:[30] transform-based compression, delta-encoding, and entropy-based compression. These techniques are depicted in Fig. 2.25.

Transform-based compression is depicted in Fig. 2.25a. After the signal is sampled, it is mathematically transformed to a domain in which the sensor data can be represented in a sparse basis. Examples of this are, for example, audio signals, which are more sparse in the frequency domain, ECG signals that are sparse in certain wavelet bases [30, 54, 76], etc. Typically, a transform is applied by a dedicated on-chip processor, followed by an optional thresholding (i.e., throwing away small, nearzero coefficients). This thresholding introduces inherently some signal loss, but this is usually acceptable for applications as long as the main signal characteristics are still maintained [30, 54].

Another class of compression algorithms relies on the fact that biosignals usually do not have a high DR and that the difference between subsequent samples is usually only a few LSBs and not the full ADC range. This characteristic is used in compression techniques such as zero-order-hold (ZOH) compression [30] or Δ-encoding [6, 112]. ZOH reduces the data rate of the system by only keeping the sample once a certain threshold in amplitude is crossed. Δ-encoding calculates the difference between samples and only transmits this difference. While these techniques may potentially reduce less data compared to the above transform-based methods, they are simple to implement in hardware and consume less power than the transform-based methods [30].

Finally, the digital data stream from the ADC can also be decimated by means of entropy-encoding techniques such as Huffman encoding or run-length encoding [54, 76]. These encoding techniques reduce the bitstream of the sensor node by

[30] Note that covering all possible compression techniques is beyond the scope of this work, nor is it the goal of this section. The goal is merely to give a notion to the reader of techniques that exist and that can be used in a biomedical application. Note also that trying to classify different types of algorithms inevitably leads to some discussion, as some techniques could be part of more than one category.

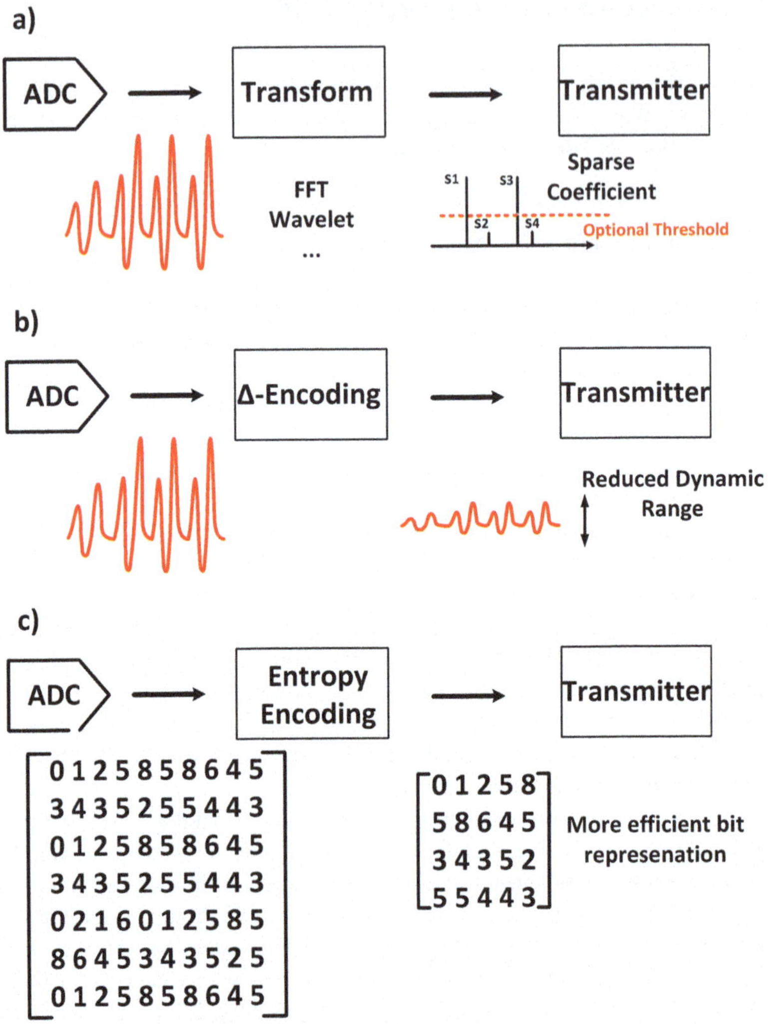

Fig. 2.25 Digital compression techniques for biosignals: (**a**) transform-based compression, (**b**) delta-encoding-based compression, and (**c**) entropy-encoding-based compression

mapping the raw sensor data to a more efficient bit representation. For example, Huffman encoding uses variable bit lengths for each symbol/output code of the ADC and assigns the least amount of bits to the most frequently occurring symbol and the most bits to the least occurring symbol. This leads to a reduction in data to transmit. Note that this type of encoding can also be applied to the output of other compression techniques [54, 56], to achieve an even larger data compression.

2.8.1.2 Digital Feature Extractor + Classifier

As discussed in the beginning of the section, if the application only requires that certain signal features are extracted or if the signal can directly be classified on chip, a significantly larger data reduction can be achieved. Certainly in systems with many recording channels, such as neural readout arrays [88], it is almost inevitable to implement such techniques, to keep the data rate of the system under control [183]. As for compression algorithms, many different feature extracting techniques and classification algorithms exist. This subsection will further discuss some examples.

Digital feature extraction is specific for the signal at hand and tries to capture those parts of the information in the signal that are relevant to the targeted classification problem. For example, neural signal features from the EEG signal are used in epilepsy classification; from the ECG signal the extracted P-QRS-T complex can be used in arrhythmia detection, etc. [31, 153, 181]. These features are very much signal-dependent. Hence, for each signal, a case-specific feature selector should be designed. The type of information that is extracted can also vary widely. For example, some features like the spectral energy in the α-, β-, and Δ-band of an EEG signal are extracted after applying an FFT to the raw sensor signal and are expensive in terms of power consumption [153]. Other features, like the time interval between two ECG R-peaks or the occurrence of the P-QRS-T complex, can be extracted in the time domain, and hence, no additional transform is required [31, 181]. This makes that the hardware cost of the feature extractor can vary widely and depends on the feature(s) needed for the classification problem.

A few different classifiers considered are depicted in Fig. 2.26. The first one is the signal-specific classifier. As the name indicates, this type of classifier is customized for a specific classification case and signal. Besides some tuning parameters, it is agnostic of the input. For ECG, an example of such an algorithm is the Pan-Tompkins algorithm, which detects the QRS-complex of an ECG signal with good accuracy [101]. In hardware, it can be implemented by a cascade of digital filters, differentiators, and integrators, followed by a threshold. Another example is [57], where the phase-synchrony biomarker from filtered recorded i-EEG data is compared to a threshold to detect the onset of a seizure. While the work reports a good power and area efficiency for the classifier, it lacked a sufficiently high sensitivity (only 75%). This highlights also the limitations of such simple classifier hardware, since they often consist of only a simple classification model and involve no training on (patient) data, it is difficult to detect more complex phenomena.

To increase the classification accuracy of the classifier and to improve the complexity that the classifier can handle, an on-chip ML algorithm can be used. The distinction is made between classical ML algorithms such as support vector machines (SVMs), decision trees, etc. and emerging neural network architectures (which will be discussed in the next paragraph). The classical ML algorithms are particularly useful to improve the detection accuracy of classification tasks compared to systems with simple filtering and thresholding, as they are able to find optimal pathways through a high-dimensional feature space by training on labeled datasets [8, 31, 136, 151]. However, careful codesign of the hardware and the

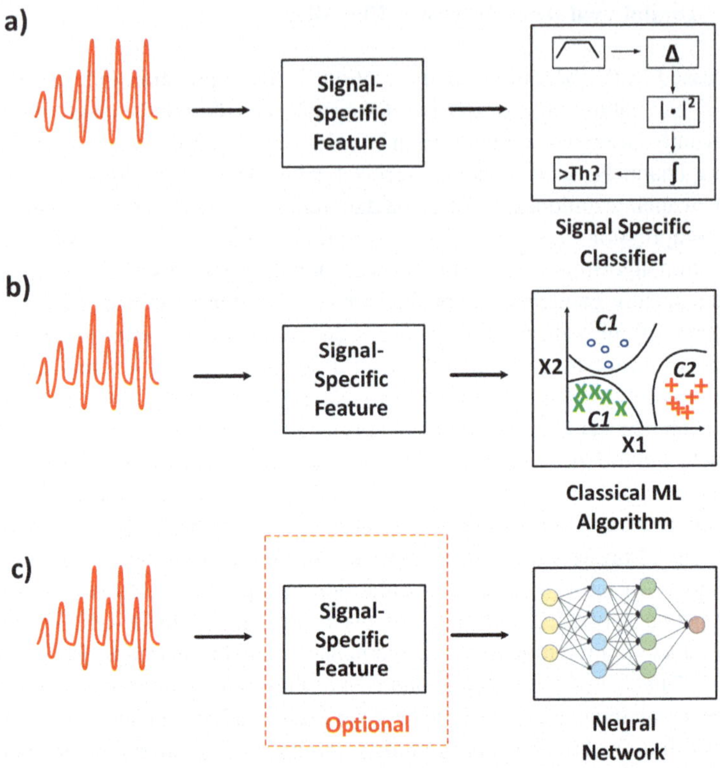

Fig. 2.26 On-chip classification algorithms: (**a**) signal-specific classifiers, (**b**) classical ML algorithm (decision trees, SVMs, etc.), and (**c**) neural networks

algorithm is necessary to implement such ML algorithms on chip, since the memory size of the on-chip SRAM required for such models can lead to excessive power consumption and excessive hardware resources being used [151]. Another downside of the classical ML techniques is that they require training on patient-specific data, and hence, the classifier should be retrained for each new patient, limiting the usability in practice [151]. An overview of some selected works that implement classical ML-based classifiers is now given. In [31] an ECG front-end with back-end DSP implements an SVM for supraventricular and ventricular arrhythmia, while only consuming 13.1 μW system power. In [136], a 256-channel SoC with AFE and back-end decision-tree algorithm and neurostimulators is presented. The decision-tree algorithm is used to classify epilepsy from i-EEG and can achieve a sensitivity of 95.6% while consuming 0.227 μJ/classification. In [8], a 16-channel SoC for EEG is presented, with two integrated linear SVMs. The two SVMs are used to analyze spectral features from the EEG signals to detect seizure, achieving a sensitivity of 95.7%, while the SoC consumes 2.73 μJ/classification.

A last type of classifier is the neural network. Since the classical ML techniques require customized features and patient-specific data, recently neural network accelerators have been explored as on-chip classifier.[31] Neural networks are able to extract their own features from the training data, and the network size can be made arbitrarily large, which can lead to a very high classification accuracy for complex classification problems [69, 151]. The caveat is that for such high accuracy, a lot of training data should be available, which is not the case for all biosignals/medical conditions [69, 151]. If few data are available, the conventional ML models might work better. Fortunately, for some classes of biosignals, such as EEG and ECG, many datasets with sufficient data points are available nowadays, and hence, neural networks can also be used successfully as classifier. For example, in [69], a neural network processor for biosignals is presented which can be used for ECG, EEG, and EMG, while consuming 30–86.7 μW depending on the signal that is processed. However, while this work can achieve a high accuracy, it also occupies an active chip area of 1.74 mm^2. In [151], a 22-channel system-on-chip (SoC) with AFE and integrated convolutional neural network (CNN) is presented, which consumes 28 μJ/classification and has an active chip area of 0.114 mm^2/channel. The advantage of this system is that the neural network only needs to be trained with *patient-unspecific* data and that, once a new patient uses the classifier, it can automatically adapt itself to the new patient. Doing this, it can achieve 86% of accuracy on actual hospital patients. As a final example, in [192] a hybrid neural network processor is proposed, which implements multiple neural network architectures, such that the best suited architecture for a specific biosignal (from ECG to APs) can be used. This processor can achieve 0.99–4.38 μJ/classification (depending on the signal and classification task) and occupies an active area of 5.06 mm^2. The question remains, however, if the area and power overhead of neural network processors, most of which is consumed by the on-chip SRAM memory to store the neural network model and its weights, is tolerable for low-power sensor systems. The power overhead of the SRAM block should be compared to the possible energy savings that the on-chip processing offers due to the reduced data rate. The possible advantages of on-chip processing thus depend on the signal and the TX protocol that is used. The acceptable area (and thus cost) overhead of the SRAM block depends on the use case of the chip (e.g., it might not be tolerable when a very cheap unit cost per system is required). Moreover, large neural network models have a high computational latency [151], which might be prohibitive for critical closed-loop applications such as neural modulators, that need to act fast to suppress, e.g., seizures [136]. To this end, possible lightweight neural networks with reduced memory size, but still with sufficient classification accuracy, could be of

[31] Important to note is that the claim of some works in literature [151] that, unlike classical ML techniques, such neural networks do not require patient-specific data is still contested. Currently, more research at the system level (patient studies) is required to fully support this claim. Typically, such studies are not performed by circuit designers, which could explain the lack of consensus within the circuit design literature.

interest. In the end of the next section, one possible candidate for such lightweight neural network, the spiking neural network (SNN), will be discussed.

2.8.2 *Mixed-Domain Processing On Chip*

While digital processing of the sensor data can achieve a large data reduction, the question can be asked if the approach in which data are first captured with a high-resolution ADC before being processed and compressed by feature extractors and classifiers is the best. Firstly, it seems wasteful to generate all these data, just to later throw most data in the bin. Secondly, the digital processor still needs to work at the Nyquist rate of the signal, and hence, while the output of the DSP algorithms may only have a low bandwidth, the DSP itself has to process the data at relatively high speeds. Hence, a push can be noted in literature to already process (part of) the signal in the analog/mixed-signal domain before or during digitization [154, 161]. Many systems in this research direction exist and are still actively being explored. In the overview below, the discussion is limited to two major techniques: compressive sensing and event-driven sampling and processing.

2.8.2.1 Compressive Sensing

The above discussion on digital compression has indicated that signals that are sampled at Nyquist rate can be transformed to sparser representations if they are transformed to another domain. Within the signal theory community, this led to the question: ***"If the signal occupies less bandwidth, in a certain domain, than dictated by the Nyquist theorem, is it possible to sample the full signal below the Nyquist rate without losing any information?"***

The compressive sensing (CS) theory deals exactly with this problem. In CS, sparse signals are sampled at a sub-Nyquist rate, to obtain direct compression when sampling the signal. This removes the need for digital compression and potentially can lower the bandwidth of the ADC, as it has to work only at a sub-Nyquist rate. The full mathematical foundations of this theory are beyond the scope of this work,[32] but some intuition is provided below. The CS theory relies on two principles: sparsity and incoherence. In the context of CS, sparsity means that the input signal (here presented as a vector) can be represented in a certain basis/domain as

$$x(t) = \sum_{n=1}^{\infty} \Psi_i(t) \cdot X_i, \tag{2.34}$$

[32] For the more mathematically brave reader, the papers from Candes [13] and Donoho [33] provide the foundations for this theory. A more layman introduction by Candes and Wakin is provided in [14].

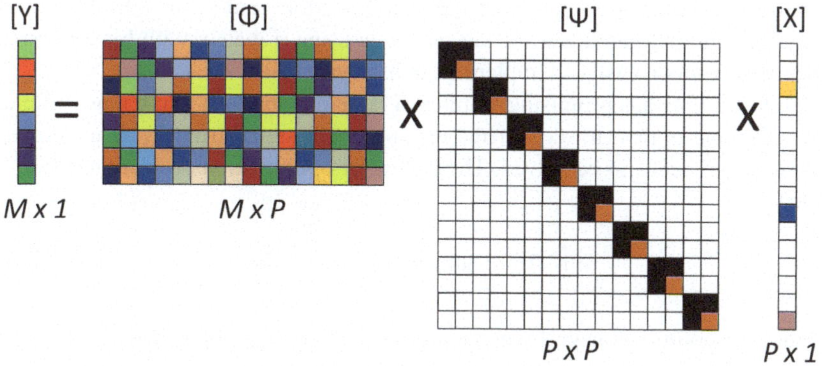

Fig. 2.27 Compressive sensing principle. The colors represent the different amplitudes of the input vector/sensing matrix, white stands for the zero value

where X_i are the coefficients of the $P \times 1$ vector X and Ψ_i are the row vectors of the basis Ψ in which $x(t)$ is sparse (see also Fig. 2.27). The vector X is said to be S-sparse, when there are only S coefficients with an amplitude larger than zero.[33] The goal of CS is now to try and sample the input signal $x(t)$ at a rate lower than P while still being able to recover the nonzero coefficients of X. As has been shown in [13, 33], this can be achieved if there is incoherence between the domain in which the signal is captured (almost always the time domain for sensor systems) and the basis in which the input signal is sparse. Intuitively, this can be understood as follows. The information in the sparse domain, which is concentrated in a few coefficients, should be spread out in the domain in which the signal is captured. For an analogy, think of a sine wave, which has one coefficient in the (single-sided) frequency domain, while being an endless CT signal in the time domain.

If the above conditions are met, as is often the case for biomedical systems, the signal can be subsampled without information loss (i.e., without aliasing). As with regular sampling, the sampling is done by means of a sensing matrix Φ,[34] which thus needs to be incoherent with Ψ. The sensing matrix dimensions are M by P, with $M << P$. The output of the product of Φ with the input vector is the compressed data stream Y (see Fig. 2.27). Thus, to implement CS in a sensor system, the sensing matrix Φ should be constructed in hardware. As the sensing matrix is crucial to the CS performance, it should be chosen to both satisfy the theoretical conditions for CS and be easy to implement in hardware. A possible candidate for such a matrix is to use a matrix with Gaussian distributed entries [14, 102]. However, generating this in the analog domain is far from trivial. Moreover, this would require a precise analog multiply-accumulate operation, which is again not easy to achieve

[33] Like in regular transform compression, coefficients near zero can be discarded to obtain a more sparse signal, at the cost of losing some signal quality.

[34] For a regular ADC, the sensing matrix just corresponds to the Dirac delta sampling pulse.

in hardware. The analog nature of the multiplication could also introduce errors [21, 102]. Another candidate for the matrix is using a matrix with binary entries, which is way much easier to implement in hardware [72, 193].

While CS thus has promising features, a large drawback of using it as compression technique is that, to recover the original input signal from Y, an underdetermined system needs to be solved. The following optimization problem needs to be solved:

$$\min_{\forall \tilde{x} \in R^n} ||\tilde{x}||_{l_1} \text{ subject to } Y = \Phi \cdot \Psi \cdot \tilde{x}. \tag{2.35}$$

Solving this optimization problem is a resource-intensive problem and hence is often performed in environments with relaxed energy constraints (e.g., the cloud) [166]. To avoid reconstructing the output signal, it is also possible to directly perform feature extraction and classification on the compressed output matrix [74, 99, 115, 137]. Herein lays also one of the strengths of analog/mixed-domain processing: The advantage of the reduced data throughput of CS is extended also to the processing circuits.

In the literature, CS has been explored in several biomedical systems. For example, a multichannel recording system for EEG is presented in [138], which can achieve a 4× compression of multichannel EEG data at a reconstruction SNR of 21.8 dB. The sensing matrix is implemented in the analog domain by means of a pseudorandom bit generator that controls which channels are summed up together. In [42], a CS sensing matrix and the multiply-accumulate operation are embedded in a SAR ADC, and the sensing matrix coefficients are multiplied by means of the CDAC of the SAR ADC. This design can achieve a compression ratio of 4 and achieves a high power efficiency by embedding the sensing matrix in the ADC, rather than implementing it with active integrators [102]. A final example of how CS can lead to system-level power savings is given in [99]. This system uses a sub-Nyquist-rate front-end for photoplethysmographic-based heart rate monitoring. The system uses the subsampled photoplethysmographic signal to extract the heart rate of a person and can achieve a compression ratio of 30×. Moreover, applying CS allows lowering the power consumption of the LED driver used in the system, as this LED driver has to work at the sampling rate of the system.

The (Non)Success of Compressive Sensing

While it is an interesting field, CS is not often used outside an academic context (with a noticeable exception for MRI applications [75]). Certainly within the circuit design community, applying CS in a system is relatively niche. In my opinion, this could be due to several reasons:

- Firstly, CS in its original conceptualization was often implemented with analog components (even though nowadays also digital CS is established)

(continued)

such as integrators, which still have to work at the Nyquist rate, hence causing some power overhead [21]. While using, e.g., passive summation methods like in [42] might overcome this, it still remains the question of what the true benefit of CS is at the circuit level.

- Secondly, to correctly implement a CS system, a good mathematical and circuit design understanding is needed. It is easy to lose performance if one of the two is lacking, which complicates the design of CS systems.
- Thirdly, while in the context of sensor systems, the reconstruction of the sensor data would occur in a base station or a cloud server, where more computational resources/relaxed energy constraints are in order, having to run (continuously) an optimization algorithm to recover the sensor data makes using CS much more complex at the system level, compared to, e.g., more simple digital compression methods such as Δ-encoding.
- Fourthly, the above three difficulties make that CS requires a lot of design effort, for possible savings in data compression that could also be achieved (or even be improved upon) by other, more simple compression methods. Hence, this may explain why CS is not often used for biomedical sensor systems in practice.

2.8.2.2 Event-Driven Sampling and Processing

While the concept of sparsity in the context of CS was rather mathematical of nature (e.g., the amount of nonzero elements in a vector), sparsity can also be used to describe the nonstationary behavior of signals such as audio, ECG, APs, etc. Such signals typically exhibit short periods in which there is information content (e.g., the QRS-complex of the ECG, the AP spike, etc.). However, the classical sensor system of Fig. 2.1 is designed to operate at the bandwidth of the system at maximum information content. In between the bursts of information, there is little information in these signals. Hence, the concept of *local* or *instantaneous bandwidth* can be defined. This is the bandwidth which is obtained by taking a short-time Fourier transform or wavelet transform of the signal fragment across a certain time interval (see Fig. 2.28). The bandwidth is then the frequency range for which the signal has a PSD value larger than a certain threshold (-100 dB/Hz in Fig. 2.28). Figure 2.28 shows for two signals, ECG and EEG, this local bandwidth, in bins of 50 ms. Clearly, for the ECG signal the maximum bandwidth occurs near the QRS-complex, while in the silent period of the signal, the local bandwidth decreases by almost a factor of 3. The EEG signal shown in Fig. 2.28 is taken from a patient experiencing a seizure [139]. Clearly the local bandwidth increases once the seizure occurs. Signals that show this type of bandwidth profile are said to be *time-sparse signals*. Exploiting the changing instantaneous bandwidth can potentially lead to large savings in both the ADCs, DSP, and TX [152], as the power of all these

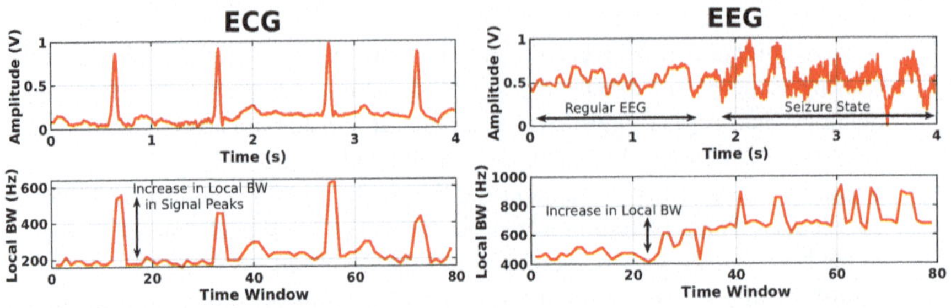

Fig. 2.28 Example of biosignals with varying signal activity (ECG and EEG) (top) and the local bandwidth calculated for these signals. The local bandwidth is calculated by applying a windowed FFT (of 50 ms) on the signals, and taking the bandwidth as the part of the spectrum with a PSD higher than −100 dB/Hz

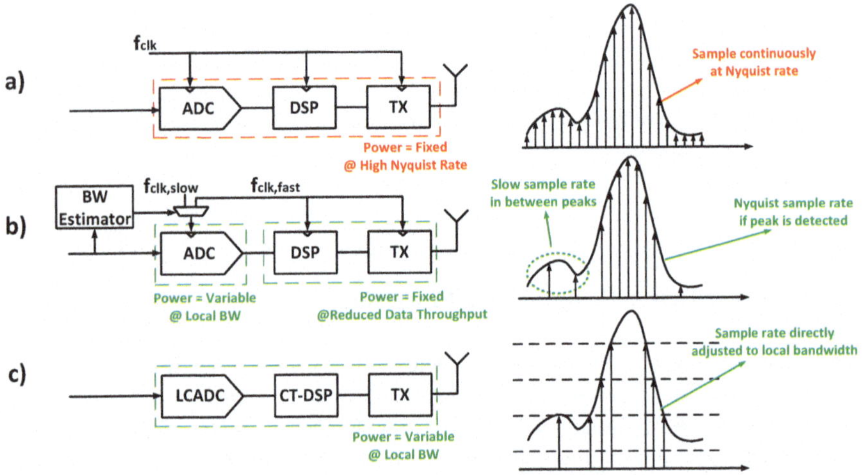

Fig. 2.29 Sensor systems for signals with varying local bandwidth: (a) Nyquist sampling (always sample at the maximum bandwidth), (b) adaptive sampling, and (c) event-driven sampling and processing

blocks depends on the sampling rate of the system. In the literature, there are two techniques that try to exploit the time-sparse characteristics of such signals: adaptive sampling and event-driven sampling and processing (see Fig. 2.29).

Adaptive sampling schemes (Fig. 2.29b) adjust the sampling rate of the ADC of the system if the local bandwidth passes a certain threshold value [59, 100, 184, 185]. This threshold value is detected via a bandwidth estimator circuit that controls the sampling rate of the ADC; it can select either the Nyquist frequency or a slower sampling frequency. When the ADC is working at the slow sampling rate, it consumes significantly less power and takes fewer samples, which then leads to a reduced processing power (the DSP has to process fewer samples) and/or TX power (fewer samples need to be transmitted) [59, 100, 184, 185]. For example, the ASIC developed in [185] has been integrated in a full wearable system and can achieve a 9× reduction in system power (compared to the same system running continuously at the Nyquist rate) when using the adaptively sampled data for processing and transmission. In [59] a SoC for ECG has been presented, and a 36% reduction in system power has been noted when it works in the adaptive sampling mode.

While adaptive sampling thus offers significant benefits at the system level, in such systems, there is still a (fast) clock operating, which makes that the whole system consumes a nearly fixed power consumption. To push the advantage of the time-sparse signals even further, event-driven sampling and processing circuits have been proposed [152] (see Fig. 2.29c). Such systems operate without a clock, i.e., asynchronously, and are only active whenever the variation in signal passes through some quantization level [152, 171]. Such a sampling scheme is called level-crossing sampling (LCS) and results in a completely signal-driven data throughput of the ADC. Level-crossing ADCs (LCADCs) therefore fully take advantage of the time-sparse characteristics and can be combined with CT signal processing techniques [126, 152] to design fully clockless SoCs. For example, in [126], a fully CT system has been presented that implemented CT digital filters that directly filter the output of an LCADC. In [24], a CT DSP could implement a filter with as good a performance as regular clocked digital filters or analog filters, while also having a factor 50× scalability (ratio of highest versus lowest power) in power consumption. In [89] and [168], CT feature extraction circuits have been implemented, which can trigger a CNN classifier once an event of interest occurs, leading to a very low system power, as the CT circuits limit the amount of time the high-performance system needs to be operational. Thus, LCADCs seem to be especially well suited for low-power wake-up systems, in which the system is only triggered once a "significant" event occurs. Only a small, low-resolution circuit is always on. By implementing these wake-up circuits with LCADCs, the need for a clock that continuously runs is removed, improving the system-level power consumption [119].

An interesting class of event-driven processing circuits is SNNs, of which an example is depicted in Fig. 2.30. Like regular neural networks, SNNs are classification algorithms that can perform classification on input data from the sensor. However, SNNs do not process data at a fixed clock rate in a frame-based manner. Rather, the output of the sensor is given as an event-driven spike train input to an SNN [36, 73, 146, 192]. When a spike is given as input to an integrate-and-fire neuron within the SNN (this can be an internal spike or a spike coming from the event-driven ADC), the membrane potential of this spiking neuron increases. Once the membrane potential exceeds a firing threshold, the membrane potential is reset, and the neuron fires a new spike to other neurons in the SNN. By monitoring the spikes coming out of the output neuron layer, classification is performed. If the input signal of an SNN is a sparse spike train, the activity in the SNN is also sparse, and the SNN potentially can lower its power to an idle state in which it consumes less power. Compared to the continuously running frame-based neural networks, SNNs

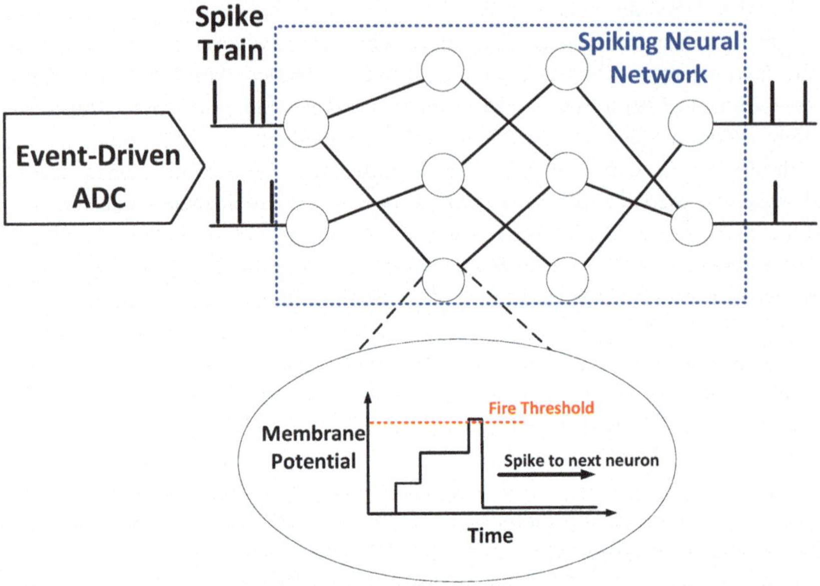

Fig. 2.30 Example of a simple spiking neural network architecture with an input layer, two hidden layers, and an output layer. Depicted in the zoomed insert is the internal functioning of a simple spiking integrate-and-fire neuron

thus provide a possible more power-efficient system [73, 146, 192].[35] A caveat is that current SNN models do not achieve the same classification accuracy for similar classification tasks as CNNs or cannot achieve more complex classification tasks compared to CNNs [122].

As event-driven LCADCs naturally fit with SNNs, there have been some systems that have co-integrated LCADCs with SNNs. In [48], two LCADCs have been integrated with an SNN and an event-driven TX stage. The system consumes 25.3 μW/channel and can reduce the required data to be transmitted off chip by 125×. In [132], a complete AFE with LCADC and back-end SNN processor has been presented for detecting high-frequency oscillations from EEG signals. The full system consists of an AFE with 8 recording channels, 32 filters, and 32 LCADCS. The filters are used to separate the inputs of the recording channels in different frequency bands; each frequency band is then converted by the LCADCs into spike trains. These spike trains are used by a 4-core SNN processor, each core consisting of 256 neurons, to detect the high-frequency oscillations from the input EEG signals.

[35] Do note that this advantage is in practice much more nuanced! As will be discussed in the last chapter of the book, SNNs are quite different from standard neural networks. Hence, comparing the two to each other is quite a difficult task. In [98] some attempt is taken to compare the standard neural networks such as the CNNs to SNNs. However, the proper benchmarking of SNNs is work in progress [186].

The total analog power consumption is 58.4 μW, while the total SNN consumes 555.6 μW. The implemented SNN can achieve an accuracy of 78%, which is state of the art for this classification task.

2.9 Conclusion

This chapter has provided a high-level system overview of typical wireless biomedical recording systems. Important design aspects have been discussed, from the electrode to the transmitter. To obtain a long battery life, the system power consumption must be minimized. An application example of a 3-lead ECG recording application has shown that most of the system power is consumed in the data transmission. To overcome this, several data reduction techniques can be used. First, digital techniques have been reviewed, which sample all data before using either compression, feature extraction, or classification in the digital domain. These techniques, while effective in reducing the data to be transmitted, can however also consume a significant amount of power themselves. Alternatively, mixed-domain techniques such as compressive sensing or event-based processing techniques can be used. These techniques already decimate the data rate of the system before or when the data are digitized, and further processing algorithms can potentially work at this reduced data rate. These techniques therefore can lower both the power of the processing and the transmission, paving the way for low-power sensor systems. While compressive sensing faces some practical limitations, event-based processing seems to offer a low system power consumption and, hence, will be explored further in this book. The next chapter will look into the basics of event-driven level-crossing ADCs and will investigate the best way to evaluate their performance.

Chapter 3
Performance Evaluation of Level-Crossing ADCs

The previous chapter has introduced various techniques to reduce the amount of data and the power consumption of data transmission in edge sensing applications. While performing feature extraction or classification in the digital domain can lead to a significant data reduction, this comes at the cost of a high power consumption of digital processing. This results in a complex system-level trade-off in which the transmitter power and the processing power need to be balanced, to achieve minimum total power consumption.

To break this trade-off, event-driven sensing and processing circuits in the mixed-signal domain can be used. These methods reduce the data throughput of the system directly when sampling and digitizing the sensor data, which can lead to both reduced power consumption for the processing circuits and/or for the transmitter. Moreover, such an event-driven sampling technique can lead to signal-dependent lower power consumption for the analog-to-digital converter (ADC) for signals that only sporadically show activity, such as the electrocardiogram (ECG) or action potentials (APs). This chapter will explore the use of event-driven level-crossing ADCs (LCADCs) for biomedical applications and will provide an in-depth comparison of such ADCs with standard Nyquist-rate ADCs and classical compression in the digital domain.

The outline of this chapter is the following. In Sect. 3.1, the LCADC topology will be introduced, its properties discussed, and examples of circuit implementations given. In Sect. 3.2, the major performance limitations of LCADCs will then be explored by means of system-level behavioral simulations in MATLAB. In Sect. 3.3, an analytical power consumption model for the classical LCADC architecture will be derived and a model for the power consumption of a Nyquist-rate ADC (the SAR ADC) will be revised. In Sect. 3.4, these models will be used to compare the LCADC with the SAR ADC, both at the ADC level and at the system level. Section 3.5 will conclude the performed analysis.

J. Van Assche, G. Gielen, *Event-Driven Circuit Architectures for Scalable and Adaptive Sensor Readout*, Analog Circuits and Signal Processing,
https://doi.org/10.1007/978-3-031-97285-0_3

3.1 The Level-Crossing ADC Architecture

3.1.1 Level-Crossing Sampling

A traditional ADC samples a signal at fixed instances in time (see Fig. 3.1a), with a sampling rate that is determined by the Nyquist frequency of the signal. Such ADCs potentially generate redundant data for time-sparse signals such as ECG or AP. In contrast, sampling signals in an event-driven manner at the rate of information or activity of the signal can alleviate this power waste and lead to more efficient sampling and processing.

An event-driven sampling scheme that shows interesting compression properties is level-crossing sampling (LCS). This technique was first introduced by Mark and Todd in 1981 [80] and later introduced again by Sayiner et al. in 1996 [125]. It is an event-driven (i.e., asynchronous) continuous-time (CT) sampling technique that only generates a sample when the input crosses a reference level, as can be seen in Fig. 3.1b. The advantage of this technique is that the sampling rate of the ADC adjusts itself dynamically to the instantaneous spectral content of the input signal. If there is a fast change in the signal, the ADC samples quickly; if the signal is static, no sample is taken. It is therefore an "event-driven" ADC. For sparse biosignals that only sporadically have a high spectral content, this technique thus has the advantage that it samples automatically at the instantaneous bandwidth of the signal [66, 152]. The LCADC thus directly compresses the signal, which reduces the amount of data generated and thus the system-level power consumption. An additional advantage of this sampling method is that the power consumption of the LCADC itself also scales with the signal activity.

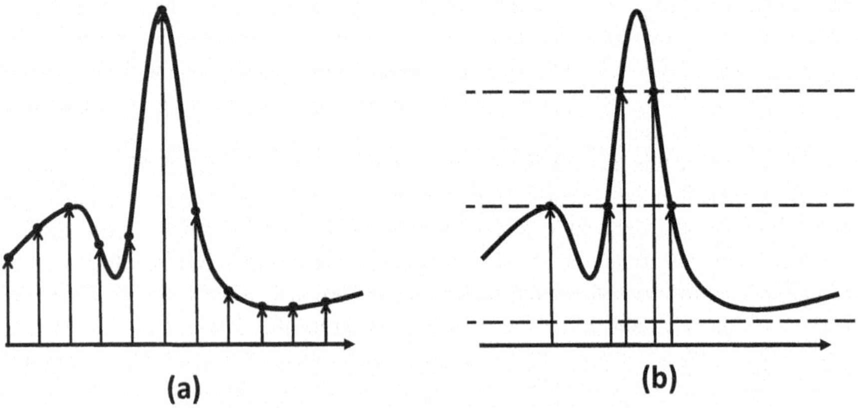

Fig. 3.1 (**a**) Uniform fixed-rate sampling scheme and (**b**) asynchronous level-crossing sampling scheme of the same input signal

3.1.2 Nyquist-Rate Sampling Versus Level-Crossing Sampling

Compared to the conventional Nyquist-rate ADC, LCADCs exhibit certain signal processing properties that are unique to this ADC topology. This subsection will briefly highlight these properties. Different implementations will be described in the next subsection.

- *Continuous-Time Sampling:* LCS is a CT sampling technique, i.e., no sample and hold stage is required, which makes that the LCADC has intrinsic anti-aliasing properties. This can lead to relaxed anti-aliasing filter (AAF) requirements, or the AAF can even be removed completely [104, 135, 178]. Note that the LCS sampling scheme should still fulfill the requirements of the Nyquist theorem to avoid aliasing, i.e., the sampling rate should be twice the bandwidth of the signal. However, due to the varying bandwidth of the signals, as long as sufficient levels are passed and the criterion is locally satisfied, no aliasing occurs [82].[1] Due to the CT sampling scheme, the LCADC is also easier to drive by the ADC input driver, as the driver is not required to provide high peak currents during the sampling phase [104, 135]. Both these properties add to the advantage of the LCADC.
- *Nonuniform sampling:* LCS is a signal-driven sampling technique: A sample is only taken at a certain level crossing, which results in a nonuniform ADC output data stream (see Fig. 3.1b). Due to this sampling method, the amplitude quantization is exact [152], and as long as the timing between two samples is preserved, the LCADC output contains no quantization errors; if this output is retimed, however, time quantization does happen. The output of the LCADC is therefore also said to be a CT digital signal [152]. This leads to an interesting observation: A signal that is amplitude sampled with LCS would only require a single amplitude quantization level to achieve a high SNR, as long as the timing in between samples is respected and the quantization level is passed sufficiently by the input signal to fulfill the Nyquist criterion. In some research works, these properties have been explored to make clockless digital filters, which work in a CT fashion [24, 126, 152]. However, while the quantization noise is absent as long as the LCADC output is not retimed, the nonuniform ADC output does introduce harmonic distortion, as the ADC turns the CT amplitude of the input signal into a discrete representation. The amount of distortion depends on the interpolation method used [178]. These properties will be analyzed further in Sect. 3.2.
- *Low latency:* Nyquist ADCs work in a clock-based manner, and the ADC output is therefore delayed with regard to the input signal. This latency between the ADC output and the input signal depends on the sampling rate of the ADC and

[1] An in-depth discussion of the concept of local bandwidth and its implications on the Nyquist theorem is beyond the scope of this work, but the interested reader is referred to [169], which provides a good starting point for the (vast) literature on this topic.

the ADC architecture. For example, a pipeline ADC has multiple clock cycles of latency, while a flash ADC has only one clock cycle of latency. LCADCs, on the other hand, sample directly once a threshold is crossed and thus directly react to a change in the signal (the LCADC will have only a small delay due to the finite bandwidth of the comparator). They therefore can exhibit a lower latency compared to a fixed-rate Nyquist ADC [164]. Nyquist ADCs can be oversampled to reduce the latency, but this comes at the expense of a higher power consumption of the ADC.

- *Decoupling of SNR and dynamic range:* A final property of LCADCs is that the dynamic range (DR) and the signal-to-noise ratio (SNR) of the ADC are decoupled.[2] Regular fixed-rate ADCs perform quantization in the amplitude, with a precision equal to the LSB of the ADC, which introduces quantization noise, and which sets a lower limit on the SNR of the ADC. The smallest input that can be detected by the ADC (and thus the DR) is also related to the quantization precision and is therefore directly related to the SNR. For LCADCs, this is not the case. Like a regular ADC, the DR is related to the resolution of the ADC quantizer. However, as discussed above, even with a low quantizer resolution, a high signal-to-noise-and-distortion ratio (SNDR) can be achieved. For example, in [148], an LCADC with 6-bit ADC resolution has an SNDR of 64 dB, which is considerably higher than the achievable DR! In other words, the LCADC can capture large (full-scale) signals very accurately but cannot detect inputs that have an amplitude smaller than the quantizer resolution. The question remains: which characteristic is more important for an application, as an ADC with limited quantizer resolution needs a large amplifier before the ADC to sufficiently amplify the input. Note that by applying techniques such as dithering, the DR can be extended greatly [163, 165], at the expense of increasing the circuit complexity (e.g., the dither extracting circuits) [165]. Moreover, the dither signal would result in an LCADC that continuously has to take a sample, which would deteriorate some of the other LCADC properties.

3.1.3 Level-Crossing ADC Implementations

Figure 3.2 depicts several ADC topologies that implement an event-driven LCS scheme. Each topology aims to exploit different properties of the LCS, depending on the application the ADC is designed for.

[2] Note that in literature this property is not well described (with some exceptions like [165]), and often SNR and DR are used interchangeably, which is wrong! For an application such as biomedical sensing, having a high SNR while having a low DR is not useful, as the input signals are never at full scale. Yet, some publications claim that, even though the DR of the LCADC is low, they get a sufficiently high SNR (at full-scale input); the question remains: which applications can tolerate this property... Probably, the confusion between DR and SNR originates from the fact that in more conventional ADC topologies, both metrics are linked to each other.

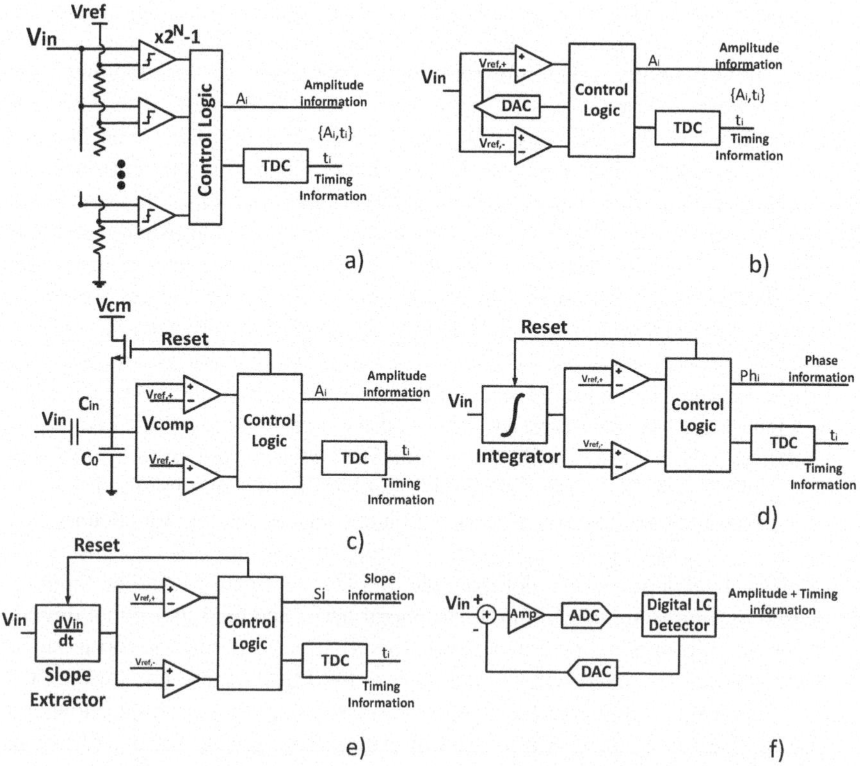

Fig. 3.2 Overview of various level-crossing ADC topologies: (**a**) flash-based level-crossing ADC, (**b**) floating-window level-crossing ADC with an N-bit DAC, (**c**) fixed-window level-crossing ADC with 1-bit DAC (implemented as in [51]), (**d**) integrating level-crossing ADC, (**e**) derivative level-crossing ADC, and (**f**) digital quasi-level-crossing delta modulator. In the figure the topologies have been simplified for clarity

- A first possible implementation is the CT flash-based LCADC (Fig. 3.2a) [174, 178]. This topology works just like a flash ADC but lacks the sample stage. If the input crosses one of the reference levels, the ADC takes a sample, and a time-to-digital converter (TDC) optionally retimes the asynchronous output. This topology, while fast, suffers from the high power consumption due to the many comparators that are required (one per quantization level) [174, 178]. In [178], this topology is explored for high-speed applications (20-MHz bandwidth) and exploits the anti-aliasing characteristics of LCS to implement a flexible, digital AAF that can remove RF blocker tones that might be present at the input of the signal chain. Another usage of this type of LCADC is in an asynchronous control loop for a low-dropout regulator, which requires a low-latency, clockless ADC [58].
- For applications that require a low power consumption, a more power-efficient delta modulator circuit like Fig. 3.2b, c can be used to implement the LCADC

[51, 66, 79, 82, 119, 144, 149, 171]. The delta modulator topology consists of only two comparators, a digital-to-analog converter (DAC), some control logic, and optionally a TDC for retiming. Two variants of the topology exist: a floating-window level-crossing topology with an N-bit DAC (Fig. 3.2b) and a fixed-window level-crossing ADC with 1-bit DAC (Fig. 3.2c). The floating-window topology (Fig. 3.2b) works as follows. The DAC outputs are two consecutive quantization levels that are spaced 1 LSB apart. Once the input signal crosses a quantization level, the comparator output becomes high and the control logic (which includes a counter) counts 1 LSB up or down, depending on the level that is crossed. In case that an asynchronous DSP is used, the samples can be processed directly. Otherwise, a TDC records the time in between two samples. After a level is crossed, the TDC outputs its current value and restarts counting, till the next level crossing. The control logic then updates the DAC according to which level has been crossed. The fixed-window topology (Fig. 3.2c) works similarly; only, instead of a DAC that is updated at each level crossing, the node V_{comp} (which tracks the input signal) is reset back to the common-mode voltage of the comparators. The two reference voltages of the comparators determine a window of 1 LSB. The delta modulator architecture is most often used in sensor or biomedical applications. For example, in [48], the floating-window topology is used to sample signals from the peripheral nerve and feed this signal directly to an on-chip spiking neural network (SNN) processor that performs signal classification. The fixed-window topology is used in [65] in a low-power ECG recording front-end. Since the fixed-window topology has a simple, low-area DAC implementation, it is also used in event-driven sensor array applications like dynamic vision sensors [67] or event-driven tactile sensors [4].

- The delta modulator topology is useful for low-bandwidth sensor applications. However, since it uses a feedback loop, the bandwidth this topology can achieve is limited. To extend the bandwidth, the integrating LCADC (Fig. 3.2d) has been proposed [103]. This topology can be implemented similarly to the delta modulator topology but integrates the signal first before the LCS scheme. To recover the original signal, the output of the ADC is differentiated. Putting an integrator before the LCS circuitry relaxes the bandwidth requirements of the delta modulator, since the output of the integrator has a slower slew rate than the input signal [103]. An additional advantage of the integrator is that thermal noise from the comparators is first-order noise shaped due to the differentiation at the end of the ADC signal path [103]. An example of this topology is implemented in [103], where this topology was used to exploit the anti-aliasing properties of LCS while having a very efficient (3–10 fJ/conv.step Walden FoM) ADC implementation. Due to the integrator (the design uses a Gm-C topology), the design can achieve a bandwidth of up to 50 MHz, significantly increasing the bandwidth compared to the regular delta modulator. In [175, 176], the integrator is implemented as a voltage-controlled oscillator, and LCS is performed in the phase domain instead of the voltage domain. These designs leverage the LCADC properties to achieve a similar performance as higher-order (fourth-order) noise-shaping CT $\Delta\Sigma$ ADCs, at much lower oversampling ratios and with only first-order noise shaping, while making use of largely digital building blocks (no OTAs are used in the design).

- Another variation of the LCADC is the derivative LCADC (Fig. 3.2e). This topology first differentiates the signal, such that the input of the delta modulator is the slope of the input signal [81], rather than the signal itself. By integrating the output of the delta modulator, the original input signal can be obtained. This ADC implementation offers, in principle, a lower possible LCADC data rate, as the slope of sparse signals changes more slowly than the actual signal [145]. However, it was shown that this is not necessarily the case for natural signals such as audio or ECG [81], likely because the derivative of such signals is not a smoothly changing function, but rather noisy, which leads to unnecessary samples in the ADC. An advantage of this LCADC is that the output is automatically linearly interpolated after integration [81]; hence, there is no need for additional interpolation hardware.
- A final possible LCADC implementation is the digital quasi-level-crossing delta modulator (Fig. 3.2f). This ADC implements LCS in the digital domain. The topology first oversamples the residue between the input signal and the previously crossed threshold level with a low-resolution (e.g., 4 bits in [164]) sub-ADC. The digitized residue is then compared to digital threshold values, and in case of a level crossing, the feedback DAC is updated. Hence, this topology is similar to the regular delta modulator topology (Fig. 3.2b, c) but benefits more from technology scaling, as a large part of the ADC is implemented in the digital domain [163, 164]. Moreover, the digital-heavy implementation increases significantly the bandwidth that can be achieved. For example, the designs in [163, 164] have a bandwidth of over 1 MHz, which is significantly higher than the bandwidth of the classical delta modulator topologies like [51, 66, 79, 82, 119, 144, 149, 171]. The topology, however, relies on a fixed-rate quantizer and requires an amplifier in the signal path that amplifies the voltage residue generated by the DAC. This results in high, fixed power consumption; hence, the quasi-level-crossing delta modulator loses the signal-dependent power advantage of the LCADC.

In conclusion, since the delta modulator topology offers a signal-dependent data rate and power consumption, it seems ideally suited for the biomedical sensing applications targeted in this book. Therefore, the remainder of this chapter will focus on the floating-window (Fig. 3.2b) and fixed-window (Fig. 3.2c) LCADCs.

3.2 Level-Crossing ADC Performance Limitations

This section will discuss the major performance limitations of the delta modulator LCADC architectures (Fig. 3.2b, c). For this analysis, a system-level behavioral MATLAB model is used, which is depicted in Fig. 3.3. The major nonidealities incorporated in this high-level model of the LCADC topology include:

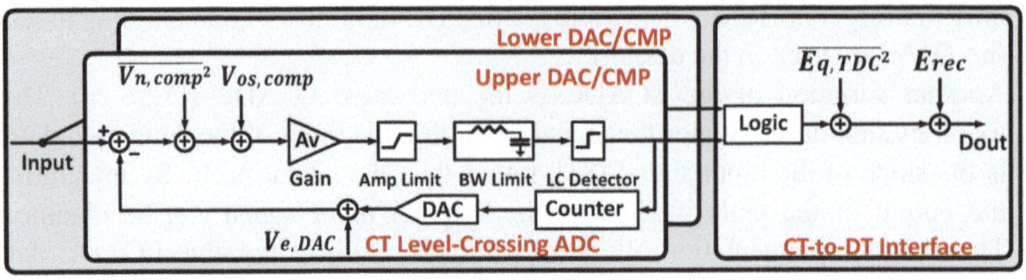

Fig. 3.3 System-level behavioral model of a CT delta modulator LCADC, with the major nonidealities included

- The quantization noise power ($\overline{E_{q,TDC}{}^2}$) and the interpolation errors (E_{rec}) of the reconstructed ADC output (indicated in the CT-to-DT Interface in Fig. 3.3)
- The mismatch errors in the DAC ($V_{e,DAC}$) and the comparator offset ($V_{os,comp}$)
- The thermal noise power of the comparator ($\overline{V_{n,comp}{}^2}$)
- The finite gain and bandwidth characteristics of the comparator, modeled as a cascade of a linear gain block Av, an amplitude-limiting block Amp Limit, and a bandwidth-limiting RC network

For illustration, the LCADCs considered have an 8-bit quantizer resolution and a 15-kHz bandwidth with a full-scale ADC input of 1 Vpp (thus, an LSB of 3.9 mV), which are representative ADC specifications for the targeted biomedical applications. The results of the analysis below are valid for any LCADC though.

3.2.1 Quantization Noise and Interpolation Errors

As LCADCs only sample when a predefined reference level is crossed (see Fig. 3.1b) [81, 152, 171, 178], such asynchronous ADCs ideally have no quantization noise, provided that the time information between samples is maintained. The amplitude quantization does introduce harmonics to the signal spectrum, which depend on the reconstruction scheme that is used to interpolate the nonuniform ADC output samples (see Fig. 3.4a) [81, 162, 178]. If real-time processing is required, typically zero-order-hold (ZOH) reconstruction is used. This reconstruction method comes "free" at the ADC output [178]; however, it introduces a large "interpolation error" and results in a large harmonic distortion. Works like [66, 79] achieve a low harmonic distortion by bringing the nonuniform ADC output off chip and by applying a high-order reconstruction method like polynomial or spline interpolation to the output. This is, however, not suited for real-time signal processing applications, as such algorithms introduce a large latency and power overhead when implemented on chip.

If the LCADC output is retimed to a synchronous clock, the retimer introduces quantization noise (see Fig. 3.4a). The resulting signal-to-quantization-noise ratio (SQNR) is given by [79]

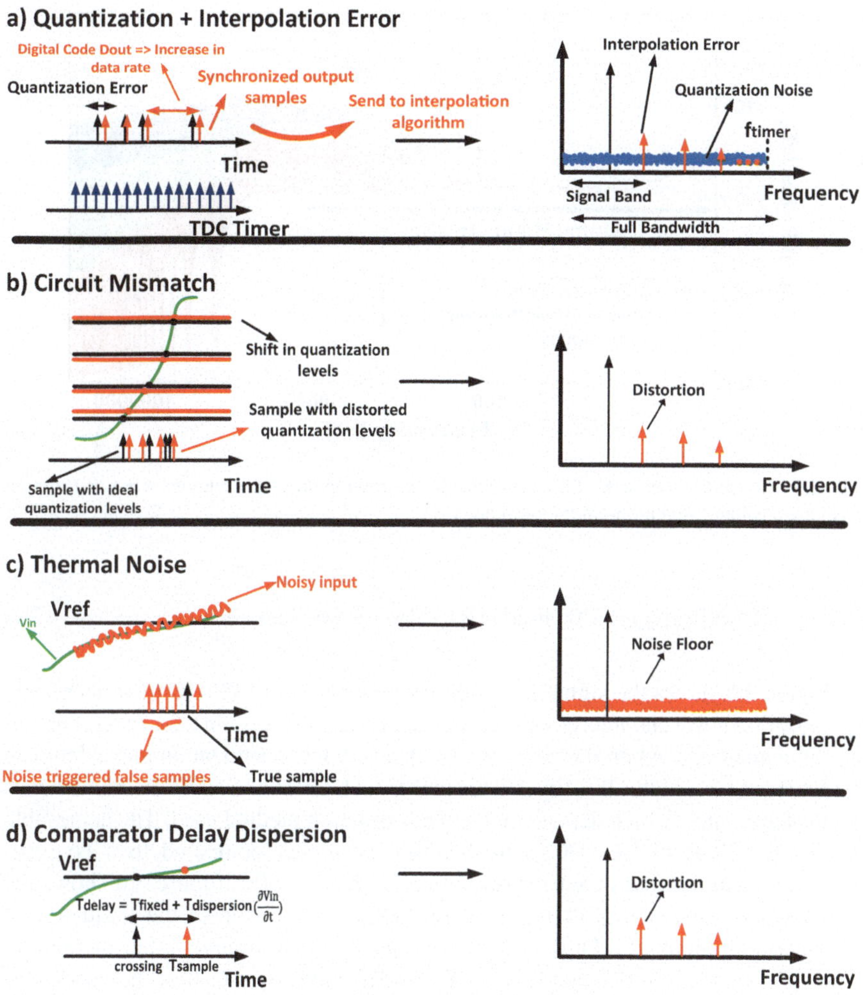

Fig. 3.4 Overview of the LCADC circuit nonidealities and noise sources, in the time and frequency domains: (**a**) quantization and interpolation errors, (**b**) circuit mismatch, (**c**) thermal noise, and (**d**) comparator delay dispersion

$$SQNR_{LCADC} = 20 \cdot \log(OSR) - 14.2, \tag{3.1}$$

where OSR is the oversampling ratio between the timer frequency f_{timer} and the input frequency f_{input}. Note that this expression gives the total SQNR, from DC to f_{timer}. However, since f_{timer} can be significantly larger than the ADC bandwidth f_{BW}, the in-band SQNR can be expressed as (adapted from [79])

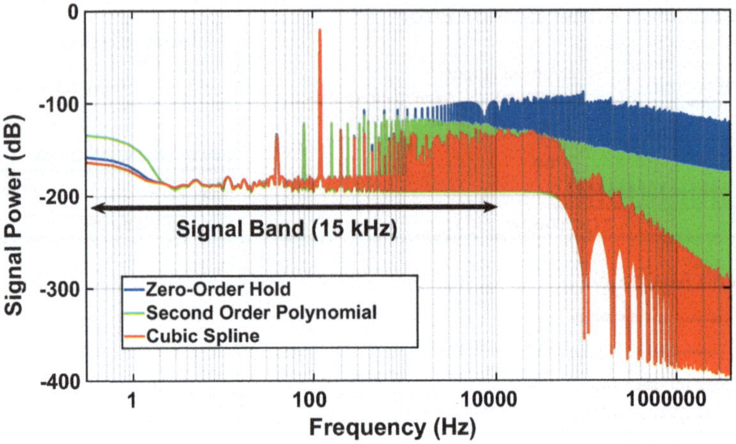

Fig. 3.5 Frequency spectrum of an 8-bit ideal level-crossing quantizer output for a 120-Hz input frequency, using different interpolation schemes

$$\mathrm{SQNR}_{\text{in-band}} = 20 \cdot \log(\mathrm{OSR}) - 14.2 + 10 \cdot \log(\frac{f_{\text{timer}}}{f_{\text{BW}}}). \tag{3.2}$$

Figure 3.5 shows the signal spectrum for an ideal 8-bit LCADC quantizer (only quantization noise and interpolation errors are taken into account), for several reconstruction methods. As predicted by Eq. (3.2), a very low quantization noise floor can be achieved for a high-OSR timer. This makes that the ADC SNDR mainly depends on the harmonics, which depend on the reconstruction method used. The achievable (in-band) SNDR of LCADCs can therefore be higher compared to a Nyquist-rate ADC with similar quantizer resolution [152, 171, 178]. Figure 3.6 shows the simulated in-band SNDR of the ideal LCADC as a function of the frequency of a single-tone input (the OSR is kept constant at an illustrative value of 66666). The ADC bandwidth is equal to 15 kHz. At low input frequencies, the harmonics caused by the interpolation method fall in the signal band, leading to a degradation of the in-band SNDR [62, 178]. Since higher-order reconstruction schemes cause less harmonics, they suffer less from such in-band SNDR degradation compared to ZOH and may achieve a higher SNDR.

3.2.2 Circuit Mismatch

Circuit nonidealities such as the comparator offset[3] and the mismatch in the DAC that generates the reference levels also introduce harmonics in the ADC output

[3] This offset is the resulting offset of the two comparators and results in a change in LSB, hence the degradation in SNDR.

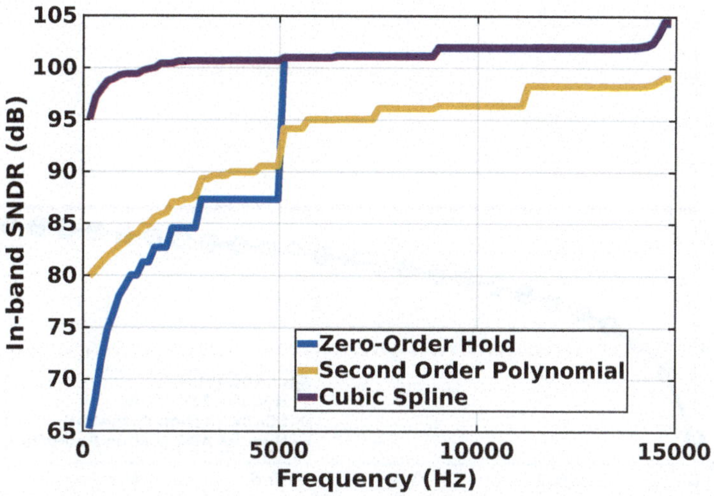

Fig. 3.6 In-band SNDR (in a 15-kHz band) for an 8-bit ideal level-crossing quantizer, as a function of the single-tone input frequency for different output reconstruction methods

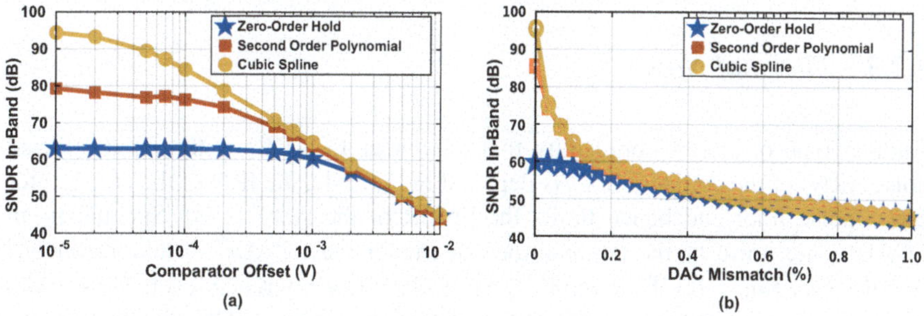

Fig. 3.7 SNDR in the 15-kHz signal band for a 120-Hz input tone as a function of (**a**) the absolute comparator offset value and (**b**) the relative static DAC mismatch (given as a percentage), for different output reconstruction schemes. The ADC LSB is equal to 3.9 mV

spectrum [48, 126, 178]. Figure 3.7 shows the in-band SNDR in a 15-kHz band for a low-frequency input tone (at 120 Hz) as a function of the comparator offset and the DAC mismatch. As can be seen, once circuit nonidealities are considered, the SNDR advantage of high-order reconstruction methods vanishes. Even if mismatch and offset would be compensated for by calibration, there is always some residual error. Hence, in practical low-power applications, ZOH reconstruction makes more sense to use.

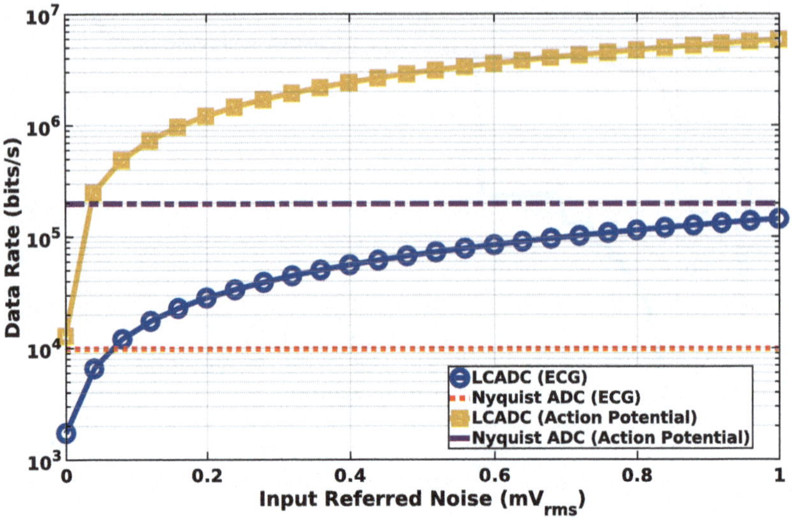

Fig. 3.8 Data rate of the LCADC as a function of the input-referred noise of the comparator for an ECG and AP signal (the Nyquist data rate is plotted as reference). The ADC LSB is equal to 3.9 mV

3.2.3 Thermal Noise

Another type of circuit nonideality that influences LCADCs is the input-referred noise (IRN) of the comparators. As depicted in Fig. 3.4c, the IRN of the comparator sets a noise floor and hence limits the SNDR of the ADC [178]. Specific to an LCADC, the noise of the comparator can trigger the LCADC to take erroneous samples (see Fig. 3.4c) if the input signal is close to a reference level. These extra erroneous samples reduce the advantage of the event-driven ADC. To illustrate the effect of the noise on the ADC data rate, Fig. 3.8 plots the ADC data rate against the IRN at the comparator input (for an 8-bit quantizer), for ECG and AP signals (taken from [37, 40, 86]), and compares it to the data rate of a Nyquist ADC. If the IRN is not sufficiently low, the excessive toggling of the comparator makes the LCADC data rate even exceed the Nyquist data rate. To counter this, in literature either comparators with hysteresis are used [66, 126, 190] or the comparators are turned off for a set amount of time after a level crossing [167, 168]; this technique is called time shielding. Both solutions, however, come with some disadvantage. Using a hysteresis introduces distortion; hence, there is a trade-off between the ADC accuracy and the amount of false samples taken. This trade-off is shown in Fig. 3.9, which plots the ADC data rate as a function of the hysteresis (Fig. 3.9a) and the in-band SNDR degradation due to the hysteresis (Fig. 3.9b). Similarly to circuit mismatch, the hysteresis nonlinearity diminishes the effect of using higher-order

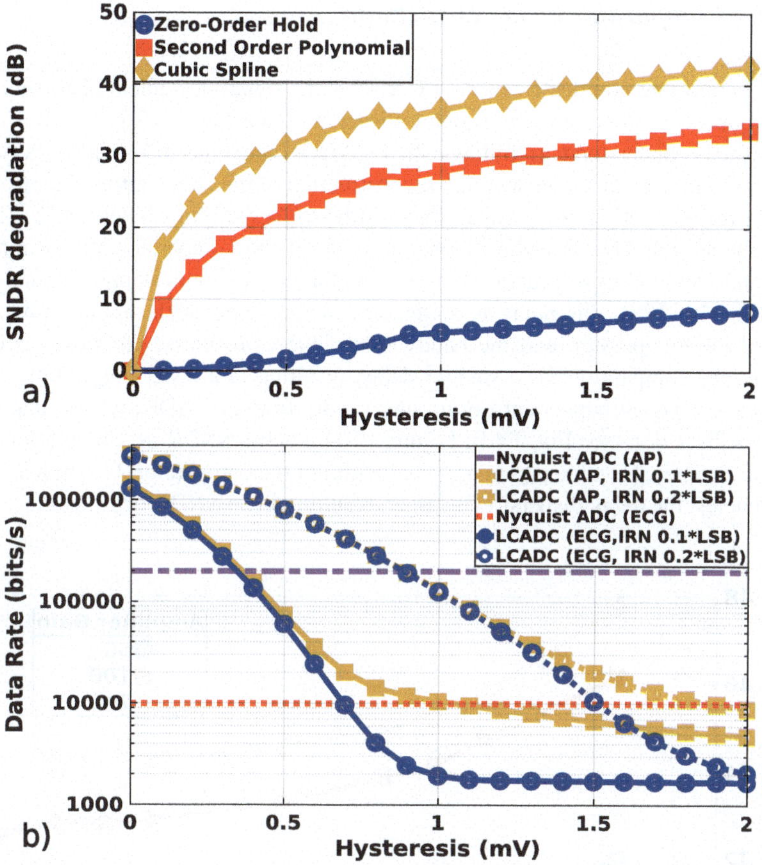

Fig. 3.9 (**a**) SNDR degradation in the 15-kHz band and (**b**) the LCADC data rate for and AP and ECG signal for different IRN values as a function of the comparator hysteresis. The ADC LSB is equal to 3.9 mV

reconstruction schemes. From the simulations, it follows that the hysteresis should be taken equal to the 3-σ value of the total integrated IRN of the comparator to avoid false level crossings. Time shielding, on the other hand, does not require hysteresis; however, the maximum time the comparators can be powered down depends on the maximum slew rate of the input signal and the ADC resolution [168]. If the comparator is still turned off once the input signal crosses a new reference level, distortion is introduced [66]. Hence, the comparator should only be turned off for a limited amount of time, after which the ADC can still take erroneous extra samples.

3.2.4 Comparator Delay Dispersion

A final important nonideality of the LCADC is the signal-dependent distortion that is introduced by the comparator delay dispersion (see Fig. 3.4d) [170, 171, 178]. This can be understood as follows. In LCADCs, the time information between samples is important for an accurate reconstruction of the ADC output. To ensure a perfect reconstruction, the comparator must be triggered immediately once a level crossing occurs. This is however only possible if the comparator has an infinite gain and bandwidth; in practice the comparator has a certain delay. As analyzed in [170, 171, 178], this delay is a function of the slope of the input signal, the gain of the comparator, and the bandwidth of the comparator. For faster-slewing signals, the comparator has a smaller delay, resulting in a faster toggle time. This introduces a signal-dependent timing uncertainty in the LCADC, which manifests itself as distortion (see Fig. 3.4d). Figure 3.10 plots the SNDR as a function of the input frequency, for a fixed gain and bandwidth of the comparator. For higher input frequencies, the delay dispersion is larger, resulting in an SNDR degradation.

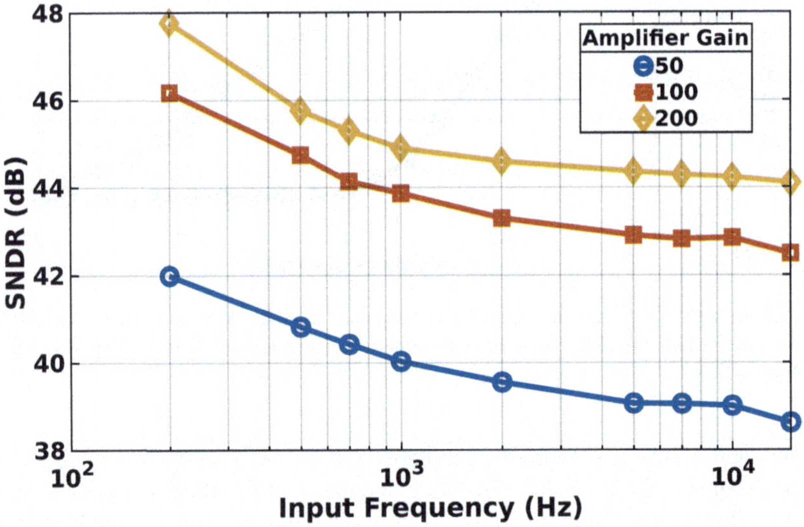

Fig. 3.10 SNDR (for ZOH reconstruction) as a function of the input frequency for an LCADC with comparator bandwidth equal to 1 MHz. The effect of different gain values of the comparator preamplifier is shown

3.3 Power Consumption Analysis of the Level-Crossing ADC and the SAR ADC

Design Tip: Power Modeling of Mixed-Signal Circuits
To evaluate mixed-signal circuits, the use of behavioral models that model the circuits in a high-level software language such as Python, MATLAB/Simulink, etc. is common practice [157], as this allows to evaluate the performance of the circuits at only a fraction of the time of circuit-level simulations. However, such models typically only model the functional behavior of the circuits (e.g., the effect of mismatch, noise, settling errors, ...), but they do not include estimations of the power consumption of these circuits. As a low power consumption is nowadays of major concern for sensor systems, a methodology has been proposed in [157] to also include high-level power models in the behavioral high-level model, which then help to optimize mixed-signal architectures toward low power consumption without having to resort to circuit simulations. The models described in this section make use of that proposed methodology.

Due to its signal-dependent power consumption, the LCADC is often proposed for use in low-power (biomedical) sensing applications for time-sparse signals. However, for this type of applications, the SAR ADC is most often used in practice, as the power efficiency of reported SAR ADCs can be significantly better (up to $100\times$) than that of LCADCs [9, 170]. To verify which ADC is actually best suited for such applications, this section will derive power models for the LCADC, such that its power efficiency can be examined and compared to the power efficiency of a SAR ADC. This section will first derive power models for both the floating-window and the fixed-window topology of Fig. 3.2b, c. Then, a model for the SAR ADC will be derived based on previous analyses made in literature.

3.3.1 Power Model for the LCADC

The main power-consuming blocks of the LCADC topology of Fig. 3.2b, c are the comparators, the DAC, the control logic, and the TDC. The total power consumption of the LCADC is thus equal to

$$P_{\text{Tot,LCADC}} = 2 \cdot P_{\text{Comp,LCADC}} + 2 \cdot P_{\text{DAC,LCADC}} + P_{\text{Logic,LCADC}} + P_{\text{TDC}}. \quad (3.3)$$

Since the LCADC needs to compare the input signal to two reference voltages [171], two comparators and two DACs are required. Expressions for the power consumption of each of the building blocks are now derived.

3.3.1.1 Continuous-Time Comparator

The comparators of the LCADC are clockless and always on, which is needed since they have to detect a level crossing instantaneously; otherwise the amplitude quantization is not exact. Hence, CT comparators that consume static DC power are required, which is indeed the case in most published LCADC implementations [66, 149, 171]. The comparators are modeled as a single-stage preamplifier with rail-to-rail output buffers, as depicted in Fig. 3.11. Note that in most circuit implementations, multiple preamplifiers are used (see Fig. 3.11a) [171, 178], but to simplify the analysis, a single amplifier stage is assumed here (see Fig. 3.11b). It is also assumed that from all blocks in the signal path (comparator, logic, and DAC), the largest delay is caused by the comparator, and thus the other blocks are assumed to have negligible delay. The ADC reference voltage and the comparator supply voltage (V_{ref} and Vdd) are also assumed to be equal.

First, an expression for the bandwidth is derived. The CT comparator's bandwidth must be sufficiently fast to detect two consecutive level crossings. If the comparator misses events, slope overload distortion is introduced, i.e., the output signal cannot track the input signal anymore and the output waveform takes the form of a sawtooth instead of the actual signal (see Fig. 3.12). This limits the ADC functionality [66, 155]. For a band-limited signal with maximum input slope $(\frac{\partial V_{in}}{\partial t})_{max}$, the minimum time between two consecutive level crossings (and thus the maximum allowable comparator delay) is given by

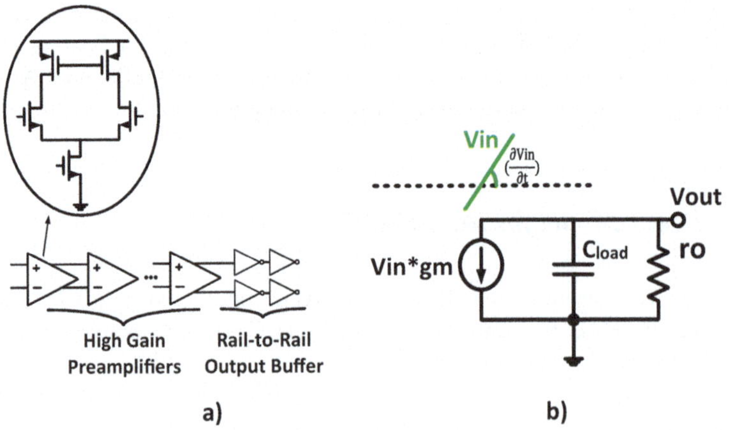

Fig. 3.11 (**a**) Opamp-based CT comparator used in the LCADC and (**b**) the small-signal equivalent single-stage model of the preamplifier used for the power model

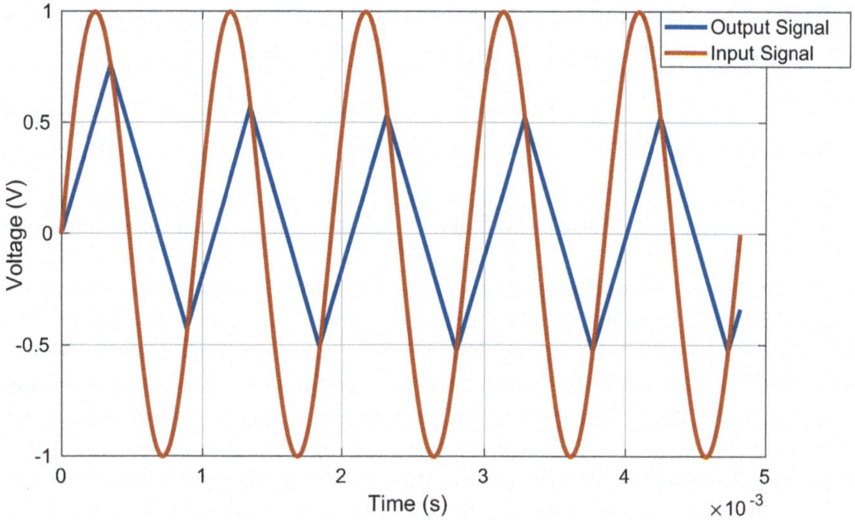

Fig. 3.12 Output voltage of an LCADC with slope overload distortion (resulting in a sawtooth output)

$$t_{\text{delay}} < \frac{\text{LSB}}{(\frac{\partial V_{\text{in}}}{\partial t})_{\max}}. \tag{3.4}$$

To calculate the comparator bandwidth, the minimum delay is set equal to three times the comparator's time constant [82, 155]. This yields the following expression for the comparator bandwidth:

$$\text{f}_{\text{BW,CMP}} = \frac{3 \cdot 2^{N-1} \cdot (\frac{\partial V_{\text{in}}}{\partial t})_{\max}}{\pi \cdot \text{Vdd}} \tag{3.5}$$

with N the ADC quantizer resolution. With the bandwidth fixed, an expression for the gain and power can now be derived to meet the different specifications. For this derivation, the comparator's input pair is assumed to be biased in weak inversion for maximal power efficiency, while still fulfilling the BW requirements. This is a reasonable assumption for low-bandwidth sensor applications implemented in a modern CMOS technology.

Typically, for medium-resolution ADCs, the (thermal) noise of the comparator of an ADC is chosen based on the LSB, such that the quantization/interpolation error is dominant. As discussed, the IRN affects the ADC's SNDR and introduces false level crossings, which are to be avoided. Therefore, the IRN is taken equal to 0.1*LSB such that the proper ADC functionality is not limited by thermal noise. Using an expression for the IRN power spectral density in weak inversion ($=\frac{4.k.T.n}{\text{gm}}$), with gm the transconductance of the input-pair transistors and n the weak-inversion slope factor, and using an expression for the equivalent noise bandwidth ($=$

$\frac{\pi . f_{BW,CMP}}{2}$), the total integrated IRN can be calculated. The power consumption of the comparator to satisfy this IRN requirement is then equal to

$$\mathrm{P_{noise}} = \frac{600 \cdot \mathrm{k} \cdot \mathrm{T} \cdot \mathrm{n} \cdot (\frac{\partial V_{\mathrm{in}}}{\partial t})_{\max} \cdot 2^{3N-1}}{\mathrm{Vdd}^2 \cdot (\frac{\mathrm{gm}}{\mathrm{Id}})_{\mathrm{WI}}}, \tag{3.6}$$

where $(\frac{\mathrm{gm}}{\mathrm{Id}})_{\mathrm{WI}}$ is the transistor efficiency in weak inversion of the input-pair transistors.

If a DC (or low-frequency) input is applied to the ADC that exceeds one of the thresholds, the comparator should toggle. Due to the finite gain of the comparator, however, the comparator only toggles once a certain sensitivity voltage ($\mathrm{V_{Sense}}$) is crossed. This effectively introduces an offset voltage for low-frequency signals, which degrades the SNDR of the ADC similarly to a "regular" offset voltage due to component mismatch. This sensitivity voltage is the minimum voltage that is required for the amplifier-stage output of the comparator (see Fig. 3.3) to reach $\mathrm{V_{th}}$, i.e., the threshold voltage of the output rail-to-rail buffers of the comparator. Thus, $\mathrm{V_{Sense}}$ can be written as

$$\mathrm{V_{Sense}} = \frac{\mathrm{V_{th}}}{\mathrm{A_v}} = \frac{\mathrm{V_{th}}}{\mathrm{gm} \cdot \mathrm{r_0}} \tag{3.7}$$

with $\mathrm{A_v}$ the DC gain of the comparator, gm the transconductance of the input transistor, and $\mathrm{r_0}$ the output impedance of the comparator preamplifier. From the bandwidth requirement and the load requirement the output impedance is calculated. The load of the preamplifier is assumed to be a minimum-sized inverter buffer, and hence, C_{load} is equal to the minimum gate capacitance of the technology C_{Logic}. The required sensitivity influences the linearity of the ADC and hence also the power consumption. As in [155], the sensitivity of the comparator is taken equal to LSB/$\sqrt{12}$. The DC power consumption of the comparator to satisfy the requirements then equals

$$\mathrm{P_{DC}} = \frac{\mathrm{V_{th}} \cdot 2^{2N} \cdot \sqrt{12} \cdot C_{\mathrm{load}} \cdot 3 \cdot (\frac{\partial V_{\mathrm{in}}}{\partial t})_{\max}}{\mathrm{Vdd} \cdot (\frac{\mathrm{gm}}{\mathrm{Id}})_{\mathrm{WI}}}. \tag{3.8}$$

A last aspect to examine is the influence of the gain on the delay dispersion performance of the ADC, which degrades the SNDR at higher frequencies. For the analysis, a 3-dB SNDR degradation is accepted. In [178], the delay dispersion was modeled, and two distinct cases were distinguished: a case in which the amplified input waveform (see Fig. 3.3) saturates before the comparator toggles and a case in which the comparator toggles before the amplified waveform saturates. In the former situation, the delay of the comparator is a function of both a ramp and DC input (after saturation, the input waveform of the RC network stays constant), while in the latter situation the delay of the comparator is the time-domain response of the RC network to a ramp function. The latter situation is assumed to be applicable for relatively low-bandwidth signals (with a small input slope) such as the biomedical signals considered in this work. The delay of the comparator can then approximately be given by [178] (assuming that $\frac{V_{ref}}{Av.(\frac{\partial V_{in}}{\partial t})_{max}} >> t_{delay,comp}$)

$$t_{delay,comp} \simeq \frac{1}{2 \cdot \pi \cdot f_{BW,CMP}} \cdot \left[1 - e^{-2 \cdot \pi \cdot f_{BW,CMP} \cdot \left(\frac{Vdd}{Av \cdot (\frac{\partial V_{in}}{\partial t})_{max}}\right)}\right]. \tag{3.9}$$

From this equation, it is clear that both the gain and the bandwidth of the comparator influence its delay. For a given bandwidth, the amount of gain required to reach the SNDR target can be calculated. This leads to the following expression for the estimated power consumption of the comparator:

$$P_{disp} = \frac{-C_{load} \cdot 9 \cdot 2^{2N} \cdot (\frac{\partial V_{in}}{\partial t})_{max}}{\ln\left(1 - \frac{3}{\sqrt{12}}\right) \cdot (\frac{gm}{Id})_{WI}}. \tag{3.10}$$

Given the three expressions Eqs. (3.6), (3.8), and (3.10), the overall power consumption of the comparator can be estimated as

$$P_{Comp,LCADC} = \max \begin{cases} P_{noise} \\ P_{DC} \\ P_{disp} \end{cases}. \tag{3.11}$$

The DC gain requirement and the delay dispersion impose that the power rises with a factor $\sim 2^{2N}$, while the noise requires the power to rise with a factor $\sim 2^{3N}$. At higher resolutions, the noise from the comparator therefore becomes the limiting factor for the power, but for practical sensor applications with medium resolutions, the stringent DC gain or dispersion delay requirements determine the power consumption of the CT comparator.

3.3.1.2 DAC

For low power consumption, the DAC implementation of the LCADC is typically a capacitive DAC (CDAC) [66, 119, 171] (see Fig. 3.13), just like for a SAR ADC. For an LCADC, the codes given to the DAC are always exactly 1 LSB apart. For simplicity, it is assumed that a simple binary-weighted DAC is used.

The energy consumption of the binary-weighted CDAC in a single conversion cycle is equal to (modified from [120])

$$E_{DAC} = V_{Ref} \cdot Q_{Tot}, \tag{3.12}$$

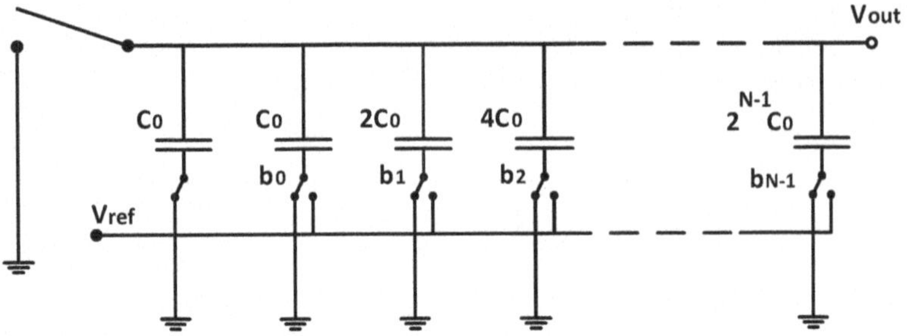

Fig. 3.13 Capacitive DAC array used in the LCADC and the SAR ADC

where Q_{Tot} is the total charge supplied by the reference voltage source V_{Ref} of the DAC (assume $V_{\mathrm{Ref}} = \mathrm{Vdd}$). For the LCADC, the DAC consumes energy each time the LCADC takes a sample. To model the power consumption of the DAC, assume that the DAC is updated at a rate of $T_{\mathrm{S,Avg}}$, the average period in between two samples taken for a certain input signal. The amount of charge supplied to the DAC of the LCADC corresponds to the code of the DAC. From [120], it follows that the total charge Q_{Tot} is equal to

$$Q_{\mathrm{Tot}} = D_{\mathrm{i}} \cdot V_{\mathrm{Ref}} \cdot C_0 \cdot \left(1 - \frac{D_{\mathrm{i}}}{2^N}\right), \tag{3.13}$$

where C_0 is the unit capacitance of the DAC and D_{i} is the code that is applied to the DAC. For the analysis, the codes D_{i} are assumed to have a uniform distribution; hence the average code equals 2^{N-1}. The estimated power consumption of the DAC for the LCADC is thus given by

$$P_{\mathrm{DAC,LCADC}} = \frac{2^{N-2} \cdot V_{\mathrm{Ref}}^2 \cdot C_0}{T_{\mathrm{S,Avg}}}. \tag{3.14}$$

As seen from the analysis in Sect. 3.2, the linearity of the DAC can degrade the effect of higher-order reconstruction schemes. The unit capacitance of the DAC should thus be chosen such that the linearity requirements of the reconstruction method are respected. For the analysis, it is assumed that the CDAC's unit capacitors are chosen such that no additional calibration of the mismatch is required. In [120], an expression for the INL of the CDAC is derived. From this expression, and assuming that the CDAC should achieve a yield of 3σ, the unit capacitor of the CDAC should be

$$C_0 = \max \begin{cases} C_{0,\mathrm{min}} \\ \left(\frac{3 \cdot \sqrt{2} \cdot \mathrm{K_C} \cdot 2^{\left(\frac{N}{2}-1\right)}}{\mathrm{INL}}\right)^2 \end{cases}, \tag{3.15}$$

where $C_{0,\min}$ is the smallest unit capacitor that can be fabricated in the technology, K_C is the matching coefficient of the capacitors in the technology, and INL is the ADC linearity expressed in LSBs. In the expression for linearity, the factor $\sqrt{2}$ is added, as the LCADC uses two DACs; thus the overall mismatch per CDAC should be a factor $\sqrt{2}$ lower to achieve the required linearity. Note also that, due to the lack of a sample and hold stage, the DAC does not need to be sized for noise, which may lead to a lower CDAC power if calibration would be used to compensate any mismatch.

For the case with a 1-bit DAC, the size of the DAC is not limited by the matching requirements, as only two external reference voltages are required for the level-crossing threshold levels.[4] However, due to charge leakage, the minimum size of the DAC capacitor is better taken larger than the minimum size capacitor of the technology [66]. A value of 50 fF is assumed for the analysis.

3.3.1.3 TDC

As timer for the LCADC, a TDC can be used to accurately measure the time between two asynchronous pulses. For the analysis a current-starved ring-oscillator TDC is used, as depicted in Fig. 3.14. There are two components of this circuit that consume power: the counters used to sample the inverter outputs and the ring oscillator itself.

The required time resolution of the timer determines the power consumption of the current-starved ring oscillator. This time resolution t_{Res} is given by

$$t_{\mathrm{Res}} = \frac{C_{\mathrm{Logic}} \cdot \mathrm{Vdd}}{I_{\mathrm{Bias}}}, \tag{3.16}$$

with I_{Bias} the bias current of the current-starved inverter used in the ring oscillator. Thus, the power consumption of the ring oscillator is

$$P_{\mathrm{RO}} = \frac{C_{\mathrm{Logic}} \cdot \mathrm{Vdd}^2}{t_{\mathrm{Res}}}. \tag{3.17}$$

Assuming that each inverter output is connected to a ripple counter (see Fig. 3.14) consisting of N_{Counter} (i.e., the resolution of the counters) D-flip-flops, the total power consumption of the counters is then

$$P_{\mathrm{Counters}} = 8 \cdot N_{\mathrm{Counter}} \cdot C_{\mathrm{Logic}} \cdot \mathrm{Vdd}^2 \cdot N_{\mathrm{Inv}} \cdot \frac{1}{t_{\mathrm{Res}} \cdot N_{\mathrm{Inv}}}, \tag{3.18}$$

[4] In a commercial chip design, of course, such reference voltages should be provided by on-chip reference buffers, and hence, the matching requirements are simply moved to another building block of the system. However, including the requirements for these other building blocks in the power model is beyond the scope of this analysis.

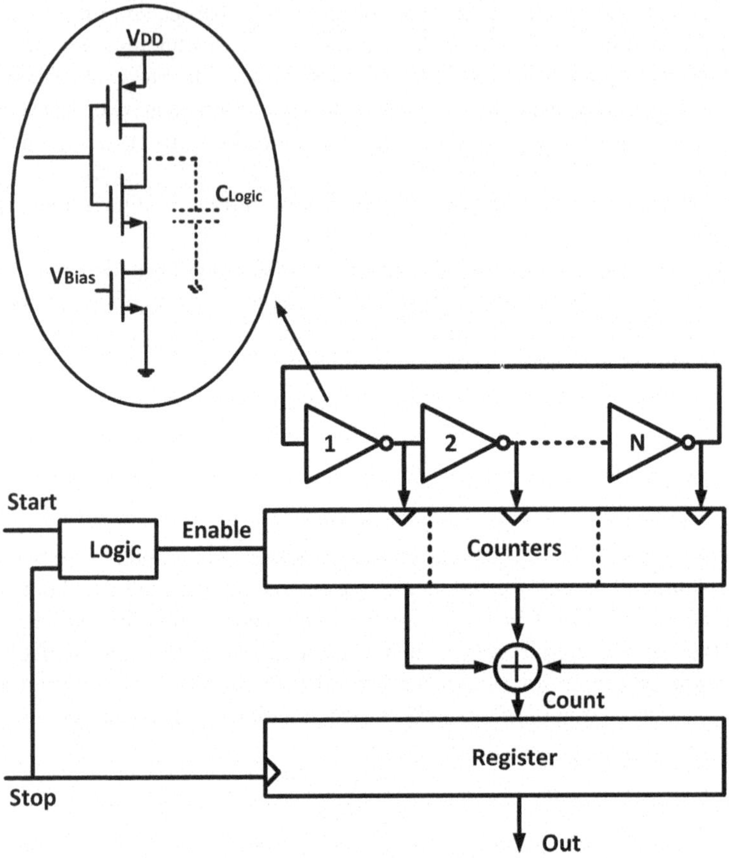

Fig. 3.14 Schematic of a ring-oscillator TDC with current-starved inverters

where a typical D-flip-flop with eight inverters is considered, and N_{Inv} is the number of inverters in the ring oscillator. This equation can be rewritten as

$$P_{\text{Counters}} = 8 \cdot C_{\text{Logic}} \cdot \text{Vdd}^2 \cdot \frac{N_{\text{Counter}}}{t_{\text{Res}}}. \tag{3.19}$$

For the TDC, there is a choice in the number of inverters N_{Inv} that is used. More inverters means that more counters have to be added (each inverter output is connected to a counter), but they can operate at a lower frequency. Likewise, fewer inverters results in less counters, but they need to operate at a higher frequency. In a first-order approximation, the number of inverters in the ring oscillator therefore does not affect the power consumption of the TDC. However, due to hardware limitations (the maximum frequency of the counters, area limitations, etc.), there is a limit on the minimum and maximum allowed number of inverters. Combining Eqs. (3.17) and (3.19) yields the total power consumption of the TDC:

$$P_{\text{TDC}} = (8 \cdot N_{\text{Counter}} + 1) \cdot \text{Vdd}^2 \cdot \frac{C_{\text{Logic}}}{t_{\text{Res}}}. \tag{3.20}$$

The counter resolution determines how long the timer keeps track of the time in between two samples. This is, however, not known a priori for most signals. To estimate the required resolution, the following approximation is used. For every class of signals, the average time the LCADC does not take a sample is taken and divided by the time resolution t_{Res}. The obtained number is rounded up to the nearest power of 2 to get a value for N_{Counter}. This gives the following estimate for N_{Counter}:

$$N_{\text{Counter}} = ceil\left[\log_2\left(\frac{T_{\text{S,Avg}}}{t_{\text{Res}}}\right)\right]. \tag{3.21}$$

For a given signal, the minimum required timing resolution can be estimated as follows. The timer should resolve between two samples, i.e., if two samples are taken, the timer resolution should be finer than the time in between those two samples. The smallest time between two samples appears when the signal slope ($\frac{\partial V_{\text{In}}}{\partial t}$) is maximal. The required time resolution is therefore equal to the maximum loop delay that is tolerated:

$$t_{\text{Res}} = \frac{V_{\text{LSB}}}{\left(\frac{\partial V_{\text{In}}}{\partial t}\right)_{\max}}. \tag{3.22}$$

3.3.1.4 LCADC Logic

The final contribution to the power consumption of the LCADC comes from the control logic. In the case of the 1-bit fixed-window LCDAC, this logic consists only of the reset logic for the DAC [66]. Assuming that only three gates are required for the reset logic (as in the design used in [156]), this power is equal to

$$P_{\text{Reset,LCADC}} = 2 \cdot 3 \cdot C_{\text{Logic}} \cdot \text{Vdd}^2 \cdot f_{\text{S,Avg}}, \tag{3.23}$$

where $f_{\text{S,Avg}}$ is the average sampling rate of the LCADC (this is the inverse of $T_{\text{S,Avg}}$). Since the logic gates used for the control logic are not inverters, a factor of 2 is added to account for the larger gate capacitance of the standard cells. For the N-bit floating-window LCADC, the same logic is needed, but an extra counter is required that keeps track of the reference levels. This counter consists of D-flip-flops, and the power of this counter is given by

$$P_{\text{Counter,LCADC}} = 8 \cdot C_{\text{Logic}} \cdot \text{Vdd}^2 \cdot N \cdot f_{\text{S,Avg}}. \tag{3.24}$$

The power consumption of the logic of the LCADC thus varies with the signal activity, but it also consumes a fixed power due to leakage. At low operating frequencies, this is a non-negligible contribution to the power consumption [188, 189]. This power is given by

$$P_{\text{Leakage,LCADC}} = (8 \cdot N + 3) \cdot \text{Vdd} \cdot \text{I}_{\text{leak}}, \tag{3.25}$$

where I_{leak} is the leakage current of a logic cell. The total power consumption of the LCADC's logic is thus equal to

$$P_{\text{Logic,LCADC}} = P_{\text{Reset,LCADC}} + P_{\text{Counter,LCADC}} + P_{\text{Leakage,LCADC}}. \tag{3.26}$$

The result from Eq. (3.26) can now be combined with Eqs. (3.11), (3.14), and (3.20) into Eq. (3.3) to obtain the estimated power consumption of the LCADC.

3.3.2 *Power Model for the SAR ADC*

Next, the power model for the SAR ADC is derived. The main power-consuming blocks are the DAC, the comparator, and the logic circuits [10, 12, 120, 188]. The total power consumption of the SAR ADC is thus equal to

$$P_{\text{Tot,SAR}} = P_{\text{DAC,SAR}} + P_{\text{Comp,SAR}} + P_{\text{Logic,SAR}}. \tag{3.27}$$

3.3.2.1 DAC

The DAC in a SAR ADC is usually a capacitive DAC (like in Fig. 3.13), since this topology only consumes dynamic power [120]. For a conventional binary-weighted capacitive DAC array, [120] derives that the power consumption is approximately equal to

$$P_{\text{DAC,SAR}} = 0.66 \cdot 2^N \cdot \left(\frac{(f_{\text{clk}} - f_{\text{Sample}}) \cdot C_0 \cdot V_{\text{Ref}}^2}{N+1} \right), \tag{3.28}$$

where N is the resolution of the ADC, f_{clk} is the clock frequency of the SAR ADC (this frequency is equal to $(N+1)\times$ the required sampling frequency f_{Sample} of the ADC), C_0 is the unit capacitance of the DAC, and V_{Ref} is the reference voltage of the DAC. The CDAC power consumption is actually code dependent (i.e., it depends on the exact input signal), but this equation assumes that the input signal has uniformly distributed input codes [120].

For the CDAC in a SAR ADC, the unit capacitors can be limited by technology, mismatch, or noise. For the mismatch limit, it is assumed that the ADC has to meet the INL requirements with 3σ yield. For the noise limit, the kT/C sampling noise power $V^2_{\text{n,kT/C}}$ is chosen as a percentage (10% in the presented analysis) of the quantization noise power. The unit capacitor of the CDAC is then equal to

$$C_0 = \max \begin{cases} C_{0,\min} \\ \left(\frac{3 \cdot K_C \cdot 2^{(\frac{N}{2}-1)}}{\text{INL}}\right)^2 \\ \frac{k \cdot T}{2^N \cdot V^2_{\text{n,kT/C}}} \end{cases} \tag{3.29}$$

with INL the required linearity of the CDAC (expressed as LSBs). Depending on the linearity requirements and the technology process used, mismatch or noise may be dominant for the DAC power consumption (as both increase proportionally to a factor $\sim 2^N$). In this analysis, it is assumed that C_0 is chosen such that no calibration for mismatch is needed. However, in practice, high-resolution designs often include calibration to relax the mismatch specifications, such that the C_0 value is chosen closer to the thermal noise requirements [45, 105, 188].

3.3.2.2 Comparator

Most SAR ADCs use dynamic latch-based comparators with a dynamic preamplifier to achieve a low power consumption [10]. This dynamic preamplifier consumes the most power, and, since it consists only of an input transistor pair and reset switches, the power consumption of this block can be modeled as two capacitors that are discharged at each comparison. Since the SAR ADC makes N comparisons per sample cycle, the power of the preamplifier is given by

$$P_{\text{Comp,SAR}} = C_{\text{load,amp}} \cdot \text{Vdd}^2 \cdot N \cdot f_{\text{Sample}}, \tag{3.30}$$

where $C_{\text{load,amp}}$ is the load capacitance of the comparator preamplifier. This comparator can be limited by noise or, at low resolutions, by the parasitic capacitance of the transistors (assume this is equal to C_{Logic}). If the comparator is noise limited, the minimum required load capacitance can be extracted from the following expression for the total integrated noise power of a latched comparator [10] (assuming that the comparator is operating in weak inversion):

$$V^2_{\text{n,DTCMP}} = \frac{2 \cdot \text{k} \cdot \text{T} \cdot n}{Av \cdot \text{C}_{\text{load,amp}}}, \tag{3.31}$$

where Av is the gain of the comparator preamplifier. A typical value for this gain is 10 [10, 188]. Thus, $C_{\text{load,amp}}$ is equal to

$$C_{\text{load,amp}} = \max \begin{cases} C_{\text{Logic}} \\ \frac{2 \cdot \text{k} \cdot \text{T} \cdot n}{Av \cdot V^2_{\text{n,DTCMP}}} \end{cases}. \tag{3.32}$$

3.3.2.3 Logic

The last power-dissipating block of the SAR ADC is the SAR logic. In [12] it is stated that these circuits consist of $(2N + 1)$ D-flip-flops. It is assumed that the D-flip-flops use eight minimal-sized inverters. The dynamic power consumption of the logic is then given as

$$P_{\text{Logic,Dyn,SAR}} = \alpha \cdot (2N + 1) \cdot 8 \cdot C_{\text{Logic}} \cdot V_{\text{DD}}^2 \cdot (f_{\text{clk}} - f_{\text{Sample}}), \tag{3.33}$$

where α is the activity factor of the digital logic, and C_{Logic} is the gate capacitance of the logic gates. α is taken equal to 0.4 [12, 188].

At low frequencies, leakage current of the logic is non-negligible. Hence, the leakage current of the logic is written as

$$P_{\text{Logic,Leakage,SAR}} = (2N + 1) \cdot 8 \cdot V_{\text{DD}} \cdot \text{I}_{\text{leak}}, \tag{3.34}$$

with I_{leak} the leakage current of an inverter. The total power consumption of the logic is then equal to

$$P_{\text{Logic,SAR}} = P_{\text{Logic,Dyn,SAR}} + P_{\text{Logic,Leakage,SAR}}. \tag{3.35}$$

Equation (3.35) can now be combined with Eqs. (3.28) and (3.30) into Eq. (3.27) to obtain the estimated power consumption of the SAR ADC.

3.4 Power Comparison of the Level-Crossing ADC and the SAR ADC

In this section the LCADC's performance and power consumption will be compared with those of the SAR ADC, both at the ADC level and at the system level.[5]

First, using the power models from the previous sections implemented in MATLAB, the power of the LCADC and the SAR ADC will be compared for the four biosignals depicted in Fig. 3.15: an ECG signal, an electromyogram (EMG) signal, an AP signal, and an electroencephalogram (EEG) signal. Following the comparison at the ADC level, the system-level performance is evaluated. For this, the LCADC is compared to the SAR ADC using two digital compression techniques: zero-order-hold (ZOH) compression and Δ-encoding. The biomedical signals used for both comparisons are taken from [37, 40, 86, 139].

[5] This analysis was first published in [155] but has been redone in this chapter. More rigorous assumptions have been made for both ADC topologies, and the effect of technology scaling has been explored further.

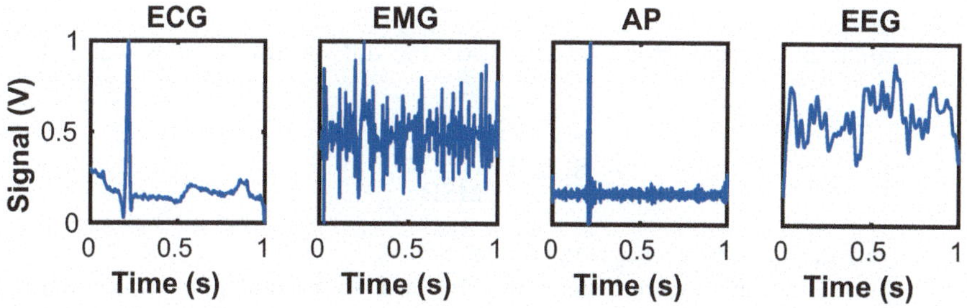

Fig. 3.15 Illustrative biomedical signals considered in the analysis, from left to right: ECG, EMG, AP, and EEG. The waveforms show the time-sparsity and the activity level of these signals

Fidelity Metrics for LCADCs

To describe how well a biomedical system captures a signal, the signal fidelity has to be measured. This fidelity largely depends on the accuracy of the ADC and the type of compression used. To characterize the accuracy of an ADC, usually the SNDR is used. However, for the SNDR, the signal assumed at the ADC input is a sine wave (which is band-limited and stationary, unlike biosignals). To generalize this metric to include generic signals, a metric called the signal-to-error ratio (SER) [81, 82] is used to characterize the "accuracy" of the ADC. The SER of a measuring system is described as the ratio of the average signal power of the input signal to the mean square error (MSE) between the original applied signal and the reconstructed signal that is derived from the compressed output signal:

$$\mathrm{MSE} = \sqrt{\left(x_{\mathrm{sig}}(t) - x_{\mathrm{rec}}(t)\right)^2}, \tag{3.36}$$

where $x_{\mathrm{sig}}(t)$ is the original applied signal, and $x_{\mathrm{rec}}(t)$ is the best fitted reconstructed signal that is compensated for gain error, DC offset, and time delay. Thus $x_{\mathrm{rec}}(t)$ is equal to

$$x_{\mathrm{rec}}(t) = a \cdot x_{\mathrm{ADC}}(t - \tau) - b \tag{3.37}$$

with $x_{\mathrm{ADC}}(t)$ the original output signal of the ADC and a, b, τ the gain, DC offset, and time delay compensation factors, respectively. The MSE contains not only the error from the ADC but also the error due to the compression and reconstruction of the signal. It is therefore suited to characterize the accuracy of the whole system and not just the ADC. Like with the SNDR, the SER can be calculated by including the error over the whole frequency band or only

(continued)

calculate the in-band error. The SER (over the full spectrum) can be defined as

$$\mathrm{SER} = \frac{P_{\mathrm{Sig,Avg}}}{\mathrm{MSE}^2}, \tag{3.38}$$

where $P_{\mathrm{Sig,Avg}}$ is the average signal power.

Since Parseval's theorem states that the MSE in the time domain is equal to that in the frequency domain, the in-band error power is defined as follows:

$$Pe_{\text{in-band}} = \int_0^{f_{\mathrm{Nyquist}}} \mathcal{F}\left(x_{\mathrm{sig}}(t) - x_{\mathrm{rec}}(t)\right)^2 df. \tag{3.39}$$

The in-band SER is then defined as

$$\mathrm{SER}_{\text{in-band}} = \frac{P_{\mathrm{Sig,Avg}}}{Pe_{\text{in-band}}}. \tag{3.40}$$

3.4.1 Power Comparison at ADC Level and at System Level

To assess the benefits of the LCADC, two comparisons with the SAR ADC are performed.

The first comparison is the standalone ADC performance of the LCADC compared to the SAR ADC. For the LCADC, both the fixed-window and the floating-window topology are considered, with and without an on-chip TDC. Using the models derived in Sect. 3.3, Eqs. (3.3), and (3.27), the power consumption needed for a required quantizer resolution will be examined for both ADC topologies. This will show for which applications which type of ADC is beneficial. Moreover, the effect of the time-sparsity of the signal and the effect of using different process technologies will be examined for the ADCs. The comparison is performed with the technology parameters given in Table 3.1. For the ADC error sources the values in Tables 3.2 and 3.3 are used. Unless explicitly stated otherwise, the values of the 40 nm technology are used for the presented results.

The second comparison will examine the system-level performance of the ADC, i.e., how well does the LCADC perform as data compression technique in a system. Since at the system level the data rate of the system is critical for the overall power consumption, the data rate that can be achieved for a certain quantizer resolution is examined for each application. For this comparison, the use cases in Fig. 3.16 are considered. The first scenario is the scenario with digital compression

Table 3.1 Technology parameters used in the power models

Technology constant	Symbol	Value (40 nm)	Value (130 nm)
Minimum load capacitance of a logic cell	C_{Logic}	0.7 fF	4.6fF
Minimum unit capacitor possible	$C_{0,\text{min}}$	0.5 fF	10fF
Matching coefficient capacitors	K_C	$0.3\ \%/\sqrt{\text{fF}}$	$0.5\ \%/\sqrt{\text{fF}}$
Threshold voltage NMOS transistor (high Vt)	V_{th}	400 mV	450 mV
Leakage current of a logic cell (Inverter)	I_{leak}	50 pA	5 pA
Transconductance over drain current ratio of transistor (in weak inversion)	$(\frac{gm}{Id})_{\text{WI}}$	$20\ \frac{1}{V}$	$25\ \frac{1}{V}$
Weak inversion slope factor	n	1	1
Nominal supply voltage	Vdd	1 V	1.2 V

Table 3.2 Error sources of the SAR ADC with digital compression

Error sources	Value
Quantization noise power ($\sigma_{e_q}^2$)	$\frac{\text{LSB}^2}{12}$
Integrated Noise power comparator ($V_{\text{n,DTCMP}}^2$)	$0.1 \cdot \sigma_{e_q}^2$
Sampling noise power ($V_{\text{n,kT/C}}^2$)	$0.1 \cdot \sigma_{e_q}^2$
INL DAC	$\frac{1}{2} \cdot \text{LSB}$
Compression error ZOH	$\Delta \cdot \sigma_{e_q}^2$

Table 3.3 Error sources of the LCADC

Error sources	Value
In-band timing error power ($E_{\text{q,TDC}}^2$)	$\frac{V_{\text{ref}} \cdot 10^{-1.42} \cdot f_{\text{BW}}}{2 \cdot (\frac{\pi}{\text{LSB}})^2 \cdot f_{\text{timer}}}$
Reconstruction error power (E_{rec}^2)	$10^{\frac{4+8 \cdot N}{10}}$
Integrated Noise power comparator ($V_{\text{n,CTCMP}}^2$)	$(0.1 \cdot \text{LSB})^2$
INL DAC	$\frac{1}{2} \cdot \text{LSB}$

applied to the output of the SAR ADC. Two compression techniques are examined: Δ-encoding (a lossless compression technique) and ZOH compression (a lossy compression technique).[6] The detailed implementation and calculation of the data rate of these algorithms can be found in Appendix A. For the LCADC, a scenario with on-chip timer (Fig. 3.16b) and with off-chip timer (Fig. 3.16c) is considered. The output of the LCADC (both the floating-window and fixed-window topology) can be represented by 2 bits. If the timer is included on chip, this adds extra power consumption and a data overhead due to the extra counter bits but has the advantage

[6] These two reference techniques have been chosen for two reasons. Firstly, both techniques also rely on the time-sparse properties of the signals to achieve data compression and hence can be used to compress the same type of signals as the ones that are suited for LCADCs. Secondly, both techniques are low-cost compression techniques, i.e., they are compression algorithms that can be implemented at a low hardware cost and a low power overhead [30]. As discussed in the previous chapter, elaborate compression algorithms based on, e.g., FFT consume a significant amount of power as well, complicating the analysis.

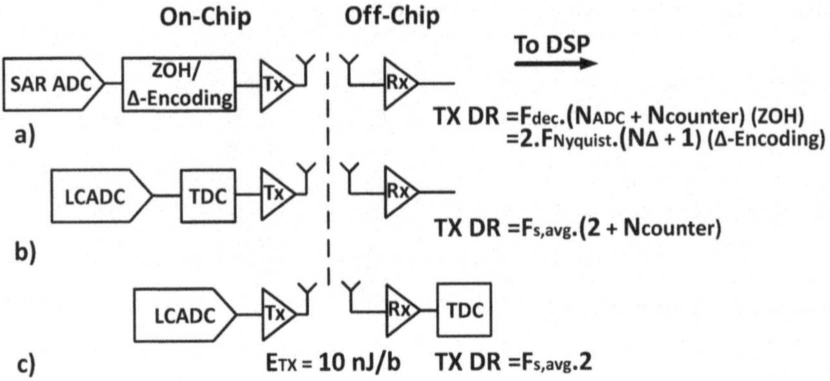

Fig. 3.16 Different use cases that are compared at the system level: (**a**) the SAR ADC with digital compression, (**b**) the LCADC with on-chip timer, and (**c**) the LCADC with event-driven transmission and with off-chip timer

Table 3.4 Nyquist frequency of the different signals used in the comparisons

ECG	500 Hz
EEG	500 Hz
EMG	2000 Hz
AP	10,000 Hz

that the data can be stored or processed further synchronously. The timer resolution of the LCADC is calculated with Eq. (3.21). For the system-level evaluation, TX transmission energy of 10 nJ/bit is assumed, which is a typical efficiency number of a Bluetooth transmitter [187].

To determine how well a compression technique reduces the data rate, a metric called the compression ratio (CR) is used. The CR is defined as [77]

$$CR = 100 \times \frac{D_{\text{Nyquist}} - D_{\text{Compressed}}}{D_{\text{Nyquist}}} \tag{3.41}$$

with D_{Nyquist} the Nyquist data rate and $D_{\text{Compressed}}$ the data rate after the signal compression. For both comparisons, the Nyquist rate of the signals can be found in Table 3.4. The ADC and system parameters that depend on the Nyquist rate of the signal, e.g., $(\frac{\partial V_{\text{in}}}{\partial t})_{\max}$, Nyquist DR, sampling frequency, etc., are calculated with the values in this table, assuming a sine input. The average DR and sampling frequency for both the LCADC and the digital compression techniques are extracted from MATLAB simulations.

3.4.2 *Detailed Results for the ECG Signal*

As an illustrative example, the results of the analysis for the ECG signal are now discussed in detail.

3.4.2.1 Comparison 1: Standalone ADC Performance

Nyquist-Rate Input Versus Sparse Input

For the ADC comparison, first the effect of a sparse input signal on the LCADC's power consumption is examined. In Fig. 3.17a, c, the LCADC power consumption is plotted as a function of the quantizer resolution, for the fixed-window and floating-window LCADC topology, respectively. Both a sine input of 500 Hz, which is equal to the bandwidth of the ECG, and a representative ECG input are simulated. Clearly, at lower resolutions (<7 bits), the fixed-window topology achieves a lower power consumption for the ECG signal compared to the sine input. For example, it achieves 4× less power for a 6-bit resolution (see Fig. 3.17b). From the power distribution plot in Fig. 3.17b, it is clear that mainly the power of the DAC and the logic scales with the signal activity, while the power consumption of the TDC and the CT comparators is independent of the signal. As these blocks also consume most of the power consumption of the LCADC, the advantage of the sparse input signal is limited. For the floating-window topology (Fig. 3.17c), the advantage of the sparse input is even less outspoken than for the fixed-window topology, mainly since the extra logic causes a higher leakage power (see Fig. 3.17d), which is independent of the input signal. Thus, while LCADCs can take advantage of the time-sparsity in signals to some extent, there are building blocks like the TDC and the comparators that, due to the always-on operation of the ADC, consume a constant amount of power and hence do not fully exploit the burst-like nature of the ECG signal.

SAR Versus LCADC Performance

Figure 3.18 depicts the power consumption of the LCADC and the SAR ADC as a function of the quantizer resolution. It becomes immediately clear that the LCADC with timer has a significant overhead in power compared to the case without timer. Hence, at the ADC level, retiming the LCADC output should only be considered if the application really requires it. Comparing the LCADC performance to the SAR ADC performance, the figure shows that up to a certain resolution, the LCADC consumes less power than the SAR ADC; for higher resolutions the SAR ADC becomes more efficient. A cross-over point above which the SAR ADC is more power efficient is clearly visible in the figure. Under the above assumptions, it is around a resolution of 6.8 bits for the fixed-window LCADC without TDC (with an ECG signal input). Note that if the LCADC has a TDC, the cross-over point shifts

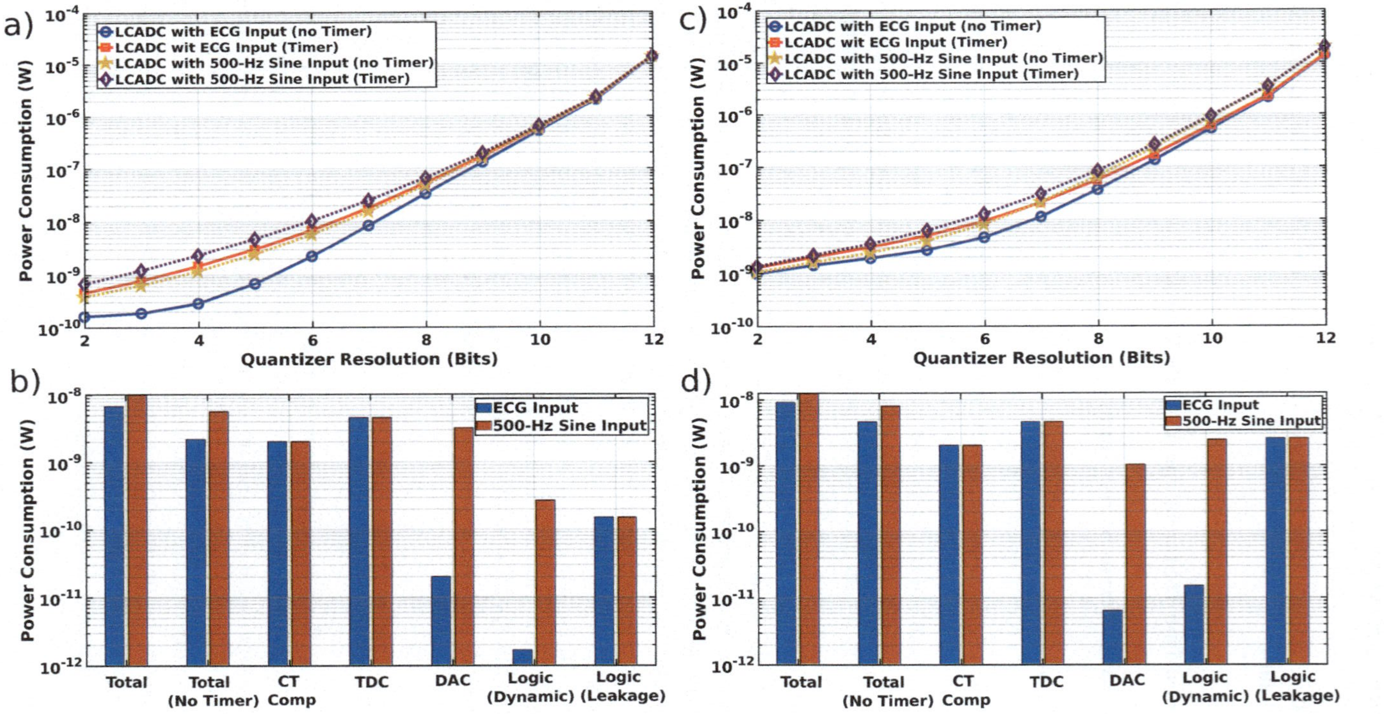

Fig. 3.17 (**a**) Power consumption of a fixed-window LCADC as a function of the quantizer resolution for a 500-Hz sine and ECG input. (**b**) Power distribution of a 6-bit fixed-window LCADC for a 500-Hz sine and ECG input. (**c**) Power consumption of a floating-window LCADC as a function of the quantizer resolution for a 500-Hz sine and ECG input. (**d**) Power distribution of a 6-bit floating-window LCADC for a 500-Hz sine and ECG input

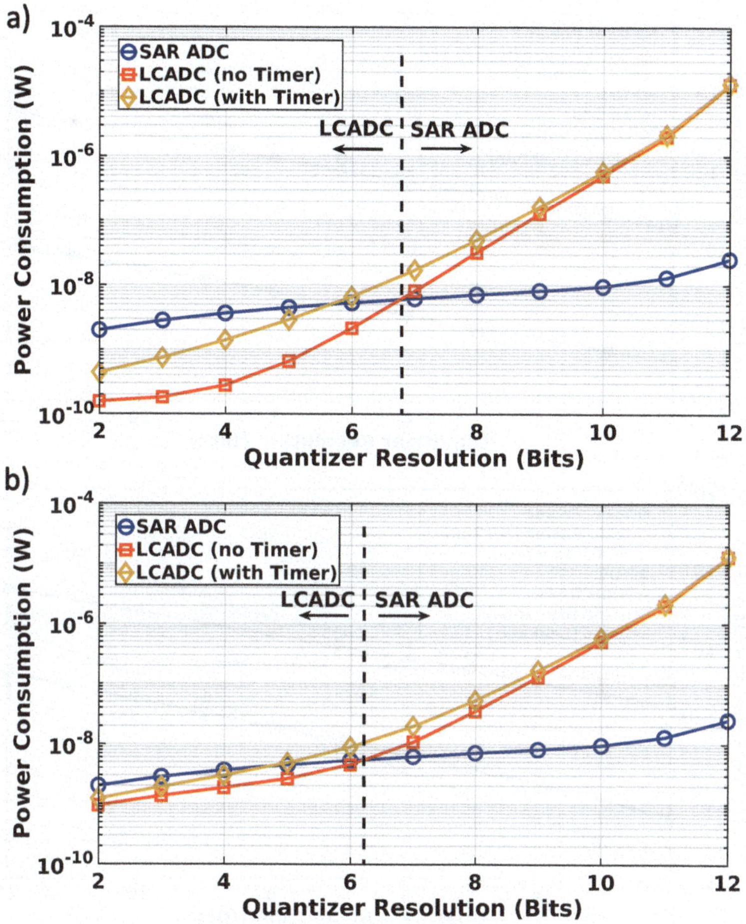

Fig. 3.18 Required power consumption as a function of the quantizer resolution for an ECG input signal for (**a**) the fixed-window and (**b**) the floating-window LCADC

downward with 1 bit in resolution. Similarly, due to the higher power consumption of the floating-window architecture, the cross-over points for that topology shift down with 0.5 bit in resolution. Thus, both LCADC topologies are a good choice for systems that only require a low-to-medium ADC resolution,[7] while the SAR ADC is more power efficient for a medium-to-high resolution. The exact cross-over points depend on the exact power consumption values in the process technology chosen.

[7] This trend seems to be confirmed by the implementations in literature, where wake-up systems with low-resolution LCADCs can achieve state-of-the-art performance [119, 168].

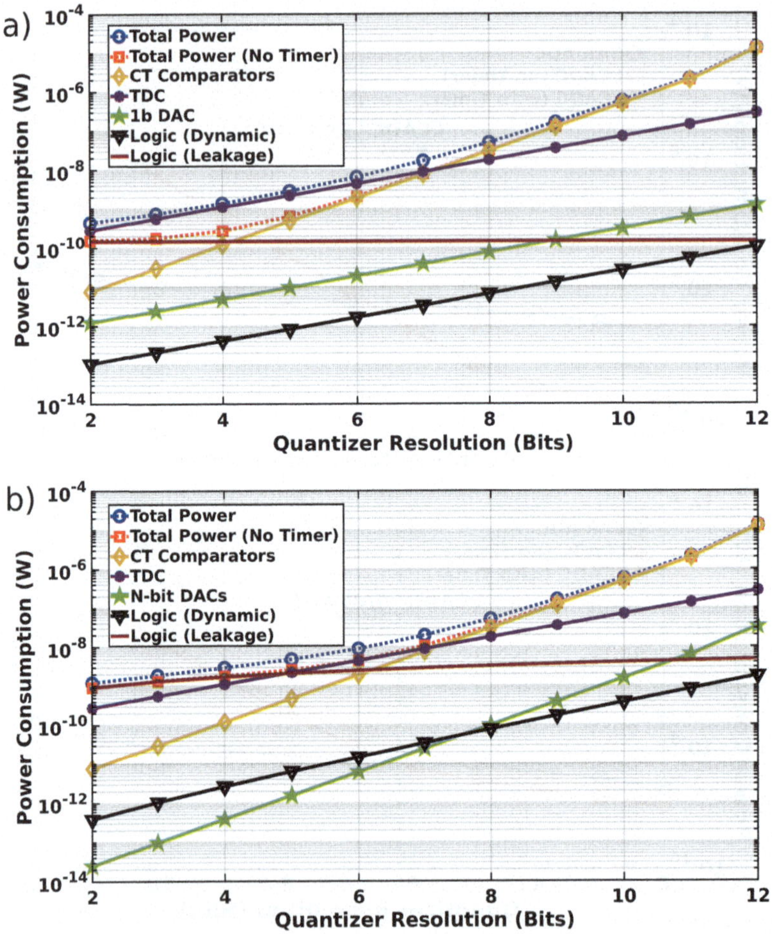

Fig. 3.19 Estimated power consumption of (**a**) the 1-bit fixed-window LCADC and (**b**) the N-bit floating-window LCADC, with and without timer as a function of the quantizer resolution for an ECG input signal

To explain the origin of this cross-over point, the power consumption of the different building blocks of the LCADC topologies is plotted as a function of the quantizer resolution in Fig. 3.19. For the LCADC without timer, the static power consumption of the two comparators is dominant from 4 bits up to higher resolutions (leakage dominates below 4 bits). For low resolutions, a comparator with low bandwidth and gain can be used, and the LCADC consumes considerably less power than the SAR ADC. However, with increasing resolution, the required bandwidth and gain of the comparator also increase (the power consumption of the LCADC increases with $\sim 2^{2N}$). For resolutions below 10 bits, the power consumption of the SAR ADC is dominated by the power of the logic, which increases with $\sim 2N^2$.

These two different trends lead to the cross-over point visible in Fig. 3.18. As the LCADC uses less logic gates compared to the SAR ADC, the actual cross-over point can shift depending on how the leakage power of the LCADC compares to that of the SAR ADC (typically, the leakage becomes dominant over dynamic power at lower ADC sampling rates). If the LCADC uses a timer, the power consumption of the timer is dominant up to a resolution of 7 bits, while at higher resolutions the power consumption of the static comparators is again dominant; the overhead in power consumption at low resolutions explains why the cross-over point shifts to lower quantizer resolutions for the LCADC with timer. If the power of the comparators and the TDC could become dynamic (e.g., proportional to the signal activity), the LCADC would become more attractive at higher resolutions.

From the power analysis, the dynamic power consumption of the logic and the DAC power of the LCADC seem to contribute up to two orders of magnitude less to the overall LCADC power consumption. However, depending on the value of the unit capacitance of the DAC and the amount of digital circuits used, the dynamic power consumption of the logic and the DAC power consumption may also become dominant at high resolutions, certainly in older technologies (which typically do not have a small unit capacitance or logic gate capacitance). This will be investigated in more detail next.

Effect of Technology Scaling

Table 3.1 lists for two technologies the various parameters used for the simulations in this section.[8] Figure 3.20 plots both for considered LCADC architectures and for the SAR ADC the power consumption as a function of the quantizer resolution, for both the values from the 40 nm technology (full lines) and the 130 nm technology (dotted lines). For the SAR ADC, going to a smaller, more advanced technology node improves the power efficiency at higher resolutions (due to the lower parasitics of the logic gates and the lower possible unit capacitor of the DAC). At lower resolutions, the efficiency of the SAR ADC degrades, since the increased leakage power in the smaller technology node dominates over the dynamic power consumption of the logic. The LCADC performance also improves with technology scaling. Firstly, the TDC's and comparator's power consumption scales with the parasitics of the technology and hence improves when using a smaller CMOS technology node. However, the comparator's power also depends on the ($\frac{\text{gm}}{\text{Id}}$) value of the technology, which tends to get worse for scaled technologies [188]. For the simulated technology parameters, the CT comparators still improve with technology scaling, as seen from the power distribution given in Fig. 3.21. This plot shows, for an 8-bit quantizer resolution, the power distribution of different building blocks

[8] These are not the exact values of such technologies, as these cannot be published. However, the two sets of parameters reflect the effects that occur when using older but more mature CMOS nodes (e.g., lower leakage, higher parasitic capacitance, etc.).

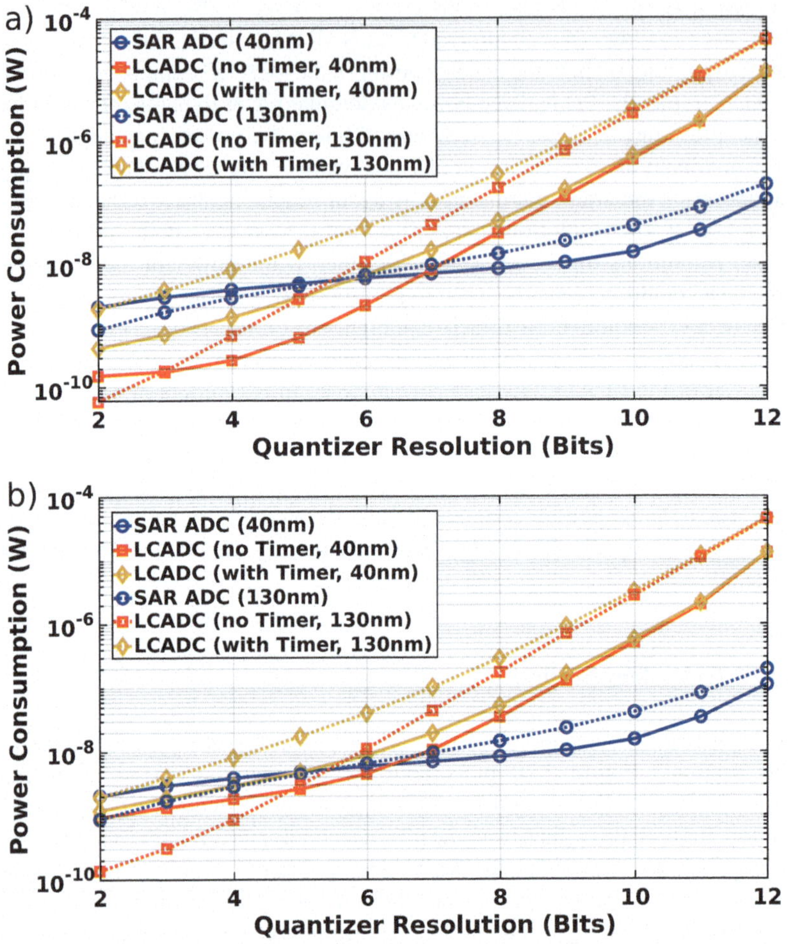

Fig. 3.20 ADC power consumption as a function of the quantizer resolution for (**a**) the fixed-window LCADC topology and (**b**) the floating-window LCADC topology, for the 40 nm (full lines) and 130 nm technology (dotted lines)

within the SAR ADC and the LCADCs. While the leakage performance gets worse in the modern technologies for all ADC topologies, the DAC and the dynamic logic power significantly decrease, due to the smaller logic gates, as well as the smaller capacitors available in the process. Thus, technology scaling helps with improving the power efficiency of the ADC. Potential leakage mitigation strategies can even further reduce the power of these ADCs [188, 189] for the targeted low-bandwidth biomedical sensing applications.

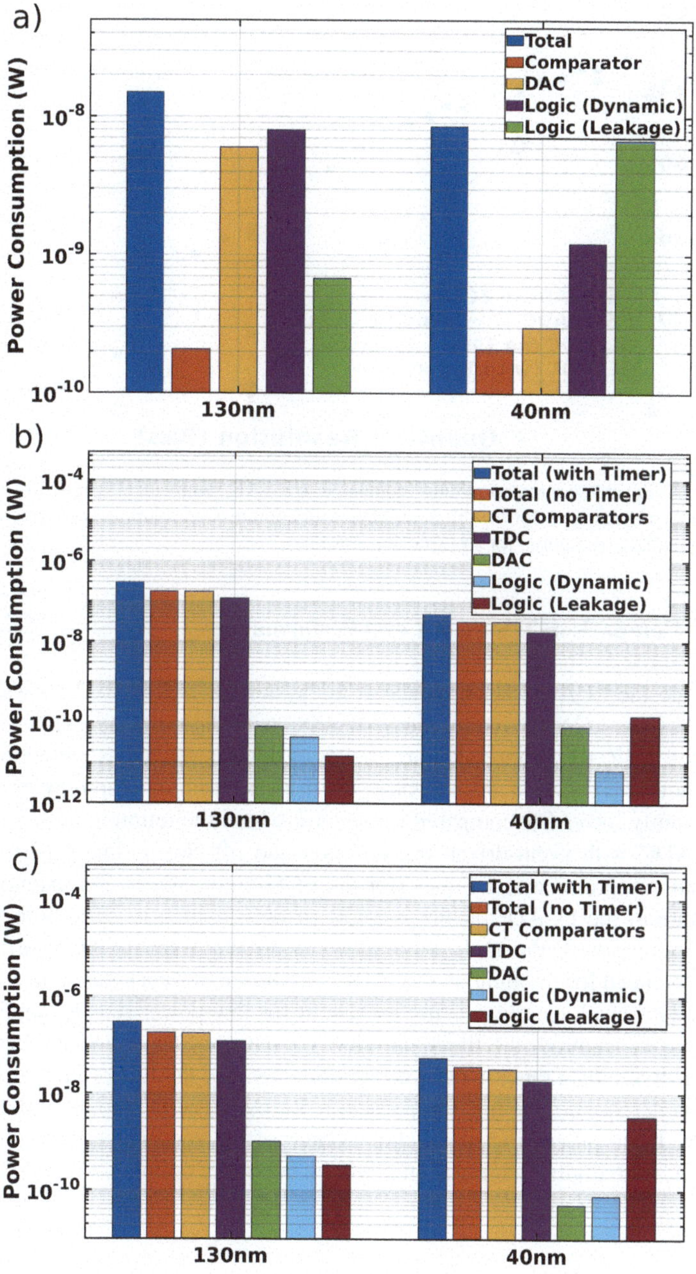

Fig. 3.21 Power consumption of the different ADC building blocks, at a 8-bit quantizer resolution in both the 40 nm and 130 nm CMOS technology, for (**a**) the SAR ADC, (**b**) the fixed-window LCADC, and (**c**) the floating-window LCADC

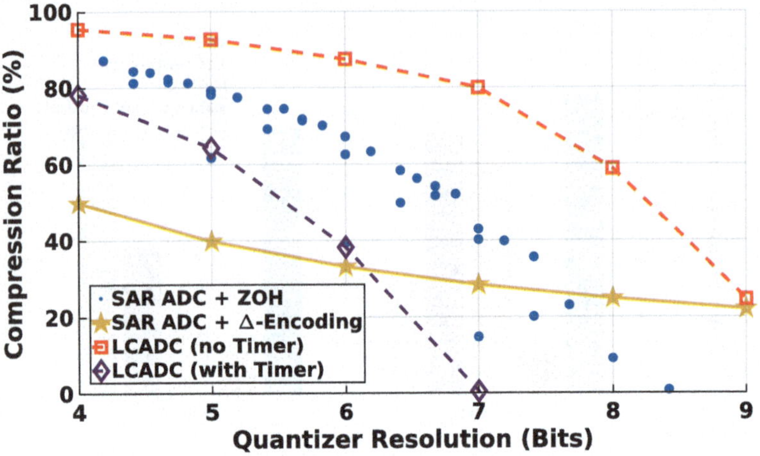

Fig. 3.22 Achievable compression ratio (compared to a SAR sampled at Nyquist frequency and without any compression), for the different ADC with compression solutions, as a function of the required quantizer resolution, for an ECG signal as input

3.4.2.2 Comparison 2: System-Level Performance

To evaluate the LCADC performance at the system level, Fig. 3.22 shows the CR achieved by the LCADC and by the SAR ADC with digital compression (different settings of the ZOH compression are shown as dots) is plotted as a function of the resolution that is required. As expected, the use case with on-chip retimer achieves a significantly lower CR compared to the case where the retiming occurs off chip. The LCADC with event-driven transmission and off-chip retiming performs the best of all compression techniques (for the ECG input signal). For example, at 6-bit resolution, the LCADC achieves a CR of 86%, which is 20% more than the ZOH compression. If the application allows to directly transmit the recording data without the need for retiming, the LCADC can thus achieve a significant advantage at the system level, clearly proving its benefit for low-power sensing applications. If retiming is required, the timer introduces a data overhead, and the benefits of the event-driven data rate are reduced significantly. For the ECG signal, if 6 bits of resolution is required, the timer introduces a 4× overhead in amount of data. Thus, when fixed-rate processing or on-chip storage of the sensor data is required, the SAR ADC with digital compression might be the better option.

3.4.3 Discussion for Other Biomedical Signals

The results for the ECG case are now generalized to the other biomedical signals considered in Fig. 3.15. Similar trends can be observed as for the ECG signal above.

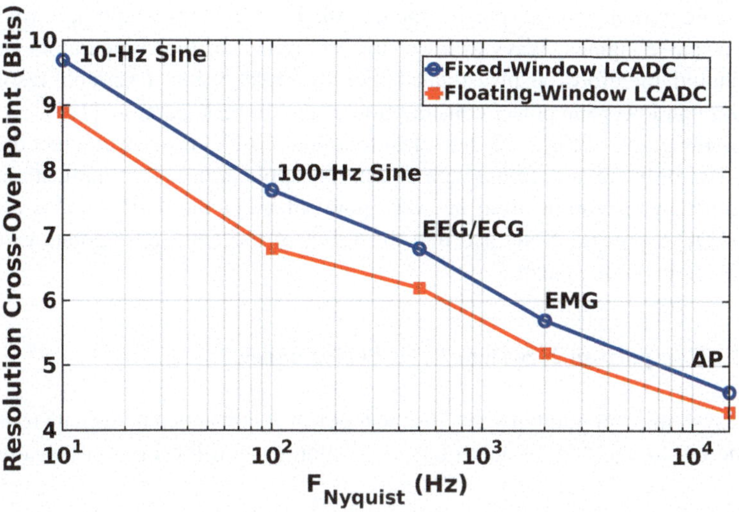

Fig. 3.23 Resolution cross-over point of the LCADC without timer and with 1-bit DAC as a function of the Nyquist frequency of the different biomedical signals

3.4.3.1 Comparison 1: Standalone ADC Performance

From the analysis of the ECG signal, it is clear that up to a certain resolution, the LCADC is more power efficient than the SAR ADC. Above this resolution cross-over point, the fixed-rate SAR ADC becomes the better option to use in an application. This cross-over point originates in the power consumption trend of both ADCs at low resolutions. For the SAR ADC, the power consumption is dominated by the logic power (either the leakage or dynamic power consumption). The LCADC power consumption is dominated by the power of the CT comparators and (if applicable) the TDC. The cross-over point (i.e., the point where these trends intersect) then depends on the Nyquist frequency, as depicted in Fig. 3.23. Indicated on this plot are the various signals considered in the analysis (as well as two low-frequency sine tones at 10 Hz and 100 Hz). The reason for this trend is that at lower frequencies, the leakage power of the SAR ADC logic starts to become dominant. At these lower frequencies, the low gain and bandwidth requirements for the LCADC's comparators result in a low power consumption. Since the LCADC also uses less logic gates than the SAR ADC, leakage is less of an issue. Hence, an improvement in cross-over resolution for LCADCs is noted for lower Nyquist frequencies.[9] Note that the cross-over point is independent of the signal activity, as the power-limiting blocks (both in the SAR ADC and in the LCADC) have fixed power consumption. This contrasts with the original idea of the LCADC that the power consumption

[9] Once dynamic power starts to dominate the power consumption of the SAR ADC logic, the cross-over point remains at a fixed resolution.

would scale with the signal sparsity and that the LCADC thus would perform better for time-sparse signals. Ways to lower the DC power of the comparators and the TDC should therefore be investigated. For example, using a variable-resolution quantizer can lower the power consumption of the two comparators [171].

From the graph of Fig. 3.23, it is clear that the LCADC is more efficient than the SAR ADC only for low-to-medium resolution applications. For applications that require a high resolution, from an ADC point of view, the SAR ADC is a better suited ADC topology. Note also that the value of the cross-over points can shift with the choice of the technology.

3.4.3.2 Comparison 2: System-Level Performance

From a system-level point of view, the most power is consumed in the data transmission and/or storage. Figure 3.24 plots the overall system-level power consumption

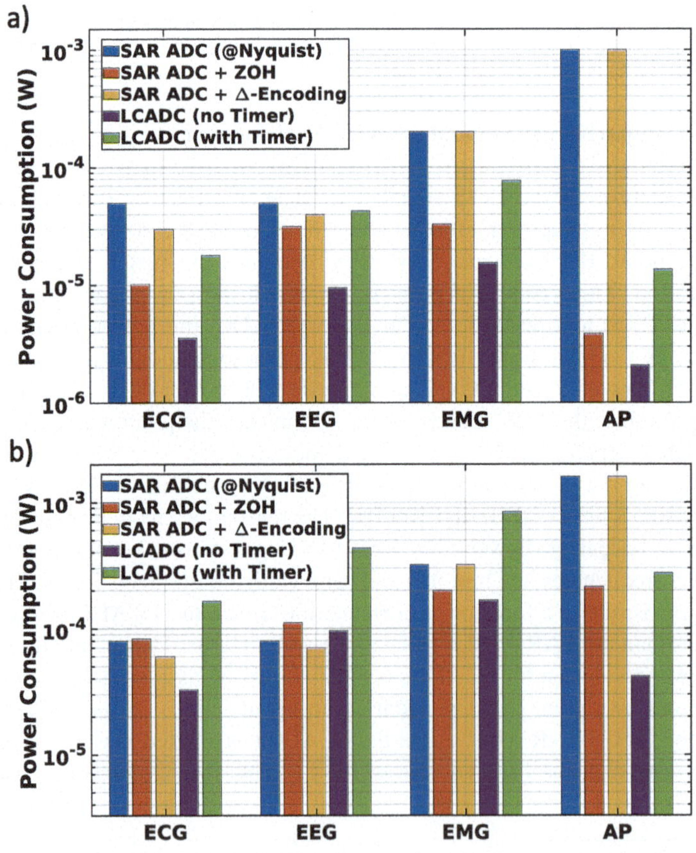

Fig. 3.24 Estimated system-level power consumption for a quantizer resolution of (**a**) 5 bits and (**b**) 8 bits for the different biomedical signals considered

for the scenarios in Fig. 3.16, for a 5- and 8-bit quantizer resolution, respectively. For a 5-bit resolution, the LCADC without timer can achieve a reduction from one (for the EMG signal) up to two orders of magnitude in system power (for the AP signal). For an 8-bit resolution, it can still reduce the system data rate significantly, up to an order of magnitude for the AP signal. The LCADC without retimer clearly outperforms the SAR ADC with Δ-encoding (which, like LCS, is also a lossless compression technique) as compression technique; for all the signals Δ-encoding achieves less compression (except for the EEG when an 8-bit resolution is required). Compared to the SAR ADC with ZOH compression, the LCADC performs better for all signals. However, for the EEG and EMG signals, the difference between both techniques is relatively small. For the use case with on-chip retimer, the LCADC performs significantly worse than the case with event-driven data transmission: Almost an order of magnitude more power is consumed due to the overhead caused by the extra bits that have to be transmitted for the retimer. Hence, in a scenario where the LCADC data are first stored before transmission, the SAR ADC + ZOH or Δ-encoding might in some cases be better. LCS can thus significantly reduce the system power, if the application allows for off-chip synchronization or if the on-chip processing can directly handle asynchronous inputs. Examples of this include SNNs [48], asynchronous wake-up circuits [119, 168], CT processors [126], etc.

3.5 Conclusion

This chapter has presented the event-driven level-crossing ADC (LCADC). After introducing the concept of level-crossing sampling (LCS) and the unique properties of this sampling scheme for ADCs, different LCADC circuit implementations have been discussed. Two implementations of the LCADC, the fixed-window and the floating-window topology, emerged as the most suited for biomedical applications, as these architectures can potentially achieve low power consumption for time-sparse signals such as ECG. Next, a system-level behavioral model of these architectures has been presented and used to evaluate the performance of the LCADC. From this analysis, it has been concluded that while LCADCs can achieve a high SNDR when using higher-order reconstruction schemes, circuit mismatches can undo some of these properties, and simple ZOH might be best to use when real-time processing is required. The effect of noise and comparator delay dispersion on the LCADC topology has also been verified. Next, power models for the power consumption of the LCADC and the SAR ADC have been presented. Using these models, for four biosignals (ECG, EEG, EMG, and AP), both the ADC performance and the performance of the entire system have been compared via MATLAB simulations. From this analysis, the LCADCs power efficiency appears to be limited by the CT comparator's and TDC's (if applicable) signal-independent power consumption. This makes that the effect of the sparse input signals, such as ECG, only has a limited advantage. Compared to the SAR ADC, the LCADC is more power efficient at lower resolutions, up to a certain resolution cross-over value which depends on the Nyquist frequency of the signal. A higher Nyquist frequency leads to a lower resolution cross-over point. Under the assumptions used in this work, for ECG and EEG signals, this cross-over in 40 nm CMOS is around a quantizer resolution of 6.8 bits. From an ADC point of view, the LCADC is therefore suited for applications where a low-to-medium resolution is needed. At the system level, the LCADC with on-chip retimer reduces the system data rate less than a SAR ADC with digital zero-order-hold (ZOH) compression. Hence, for use cases where synchronous further processing is used or where the LCADC data need to be stored, the classical Nyquist ADC with digital compression may be better to use. Without retimer, on the other hand, the LCADC proves to be more efficient at the system level: For a 5-bit resolution a system-level power consumption reduction of up to two orders of magnitude is achieved due to the LCADC's high data compression ratio, and for an 8-bit resolution, it can still achieve an order of magnitude less power. For the biosignals considered (ECG, EEG, EMG, and AP), LCADCs therefore offer a significant advantage over digital compression techniques at system level. Considering this significant power consumption reduction at system level, it can be concluded that event-driven conversion and compression schemes like LCS show great potential for use in biomedical monitoring systems where the data need to be processed or transmitted directly. The insights of this chapter will be used in the following chapters to improve the LCADC architecture, further increase its power efficiency (also for higher resolutions), and hence expand its usage potential in biomedical applications.

Chapter 4
A 10.4-ENOB Level-Crossing ADC with Adaptive Clocking Strategy

The previous chapter has analyzed the standard level-crossing analog-to-digital converter (LCADC) architecture and found that the topology can offer a significant system-level advantage over classical digital compression. However, at the ADC level, the power consumption of the continuous-time (CT) comparator limits the ADC power efficiency for higher ADC quantizer resolutions (>7 bits). The analysis of the previous chapter has been supported by various implementations of LCADCs from literature, which typically obtain a figure of merit (FoM) which can be up to two orders of magnitude larger than the best ADCs reported (which are mainly SAR ADCs). This makes that in systems with LCADCs [48, 132], the LCADC becomes the power-limiting building block. Moreover, as the previous chapter showed, the LCADC architecture becomes less attractive for higher resolutions. Previously reported LCADCs also did not achieve an effective number of bits (ENOBs) that is sufficiently high for sensing applications, and the highest resolution reported in Table 4.1 is below 10 ENOBs. The last chapter has also concluded that once a timer circuit is required to synchronize the asynchronous LCADC output—which is required for applications that have to work in a synchronous fashion—the LCADC power efficiency is further degraded. The timer also increases the output data rate of the LCADC, undoing part of the system-level benefits.

To address all these shortcomings, this chapter will present in detail an LCADC topology that uses an adaptive clocking strategy. It overcomes the limitations of previous asynchronous LCADCs by means of the following innovations:

- Power-efficient and highly accurate clocked comparators are introduced to reduce the ADC's power consumption
- An on-chip clock generator generates multiple sample clocks for the ADC to enable adaptive clocking
- Event logic dynamically adjusts the instantaneous sample clock of the ADC to the signal activity, limiting the power waste

J. Van Assche, G. Gielen, *Event-Driven Circuit Architectures for Scalable and Adaptive Sensor Readout*, Analog Circuits and Signal Processing,
https://doi.org/10.1007/978-3-031-97285-0_4

The chapter is structured as follows. First, the new adaptive clocking LCADC architecture will be introduced in Sect. 4.1. In Sect. 4.2, the new architecture will then be compared to the standard CT LCADC that was discussed in the previous chapter. Next, Sect. 4.3 will describe the detailed circuit implementation of the proposed design, and Sect. 4.4 will present the measurement results of a prototype chip, fabricated in a 40 nm CMOS technology. Section 4.5 will conclude the chapter.

4.1 A Level-Crossing ADC with Adaptive Clocking

4.1.1 The ADC Architecture

The power consumption of the CT comparators has been identified in the previous chapter as being the bottleneck of the LCADC architecture. In contrast, clocked discrete-time (DT) comparators are very efficient [79, 164], hence their use in most (clocked) data converters. Clocked comparators usually consist of a (dynamic) preamplifier and a latch that toggles (see Fig. 4.1). In [79, 164], instead of CT comparators, a clocked comparator (or sub-ADC) was used in the LCADC, which reduced the power consumption of the architecture. However, while a clocked comparator can increase the LCADC power efficiency (at the cost of introducing quantization noise), it still does not take full advantage of the time-sparse signal input, as it consumes a fixed power consumption determined by the fixed clock rate. To not miss any level crossings (and to avoid slope overload distortion, as discussed in Sect. 3.3.1.1), it can be derived that the comparator must be clocked at a frequency that is equal to the inverse of the maximum allowable loop delay, Eq. (3.4):

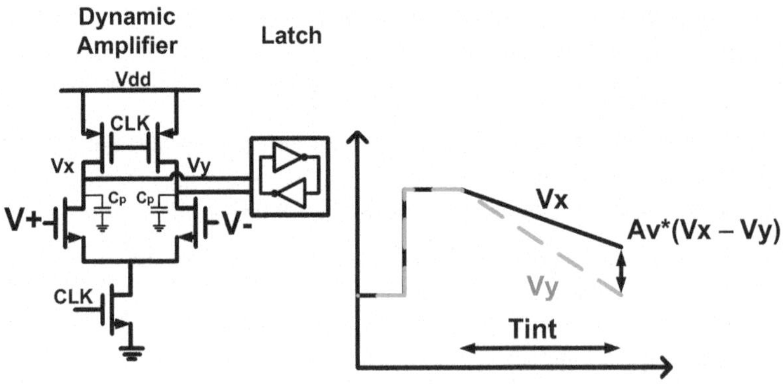

Fig. 4.1 Discrete-time comparator, with an integrating dynamic amplifier and (regenerative) latch. Shown are the output waveforms of the output nodes of the dynamic amplifier

$$f_{\text{CLK,max}} = \frac{(\frac{\partial V_{\text{in}}}{\partial t})_{\max}}{\text{LSB}}. \quad (4.1)$$

From this expression, an interesting observation can be made. To avoid slope overload, the clock rate at any moment in time should only be fast enough to track the instantaneous input signal. For time-sparse signals, the bandwidth can vary instantaneously from Nyquist rate to almost DC. With varying frequency, the input slope of the signal varies as well. Thus, the worst-case clock frequency determined from Eq. (4.1) is only required when a signal with maximum input slope is applied.

Therefore, we propose an LCADC architecture with adaptive clocking strategy (see the full architecture in Fig. 4.2). The detailed timing diagram is also shown in Fig. 4.3b. From a master clock running at twice the maximum clock rate determined from Eq. (4.1), two clocks are synthesized (CLK_DAC and CLK_COMP), which are the sample clock and the clock for the DT comparator, respectively. At each sample cycle, the input signal is sampled on the digital-to-analog converter's (DAC's) capacitors (P1), the DAC subtracts the previous code from the current input sample (P2), and the residue voltage is then compared by the comparators to two fixed references (P3 and P4). If a level crossing occurs, the counter after the comparators is updated. Depending on the signal activity, the event logic (implemented on an off-chip FPGA in our prototype implementation later on) can adaptively change the two clock signals, CLK_DAC and CLK_COMP. Since the DT comparator's power consumption scales with the clock frequency (see Fig. 4.3a), this adaptive changing of the clock leads to a lower comparator power consumption for low-activity signals such as electrocardiogram (ECG). Adapting the clock rate of the LCADC does not degrade the overall SQNR, as the OSR of the ADC remains constant. The in-band SQNR of the LCADC architecture, however, is slightly degraded and is given by (modified from [79])

$$\text{SQNR}_{\text{in-band}} = 20 \cdot \log(\text{OSR}) - 14.2 + 10 \cdot \log\left(\frac{\alpha \cdot f_{\text{CLK,max}}}{f_{\text{BW,ADC}}}\right), \quad (4.2)$$

where $\alpha \cdot f_{\text{CLK,max}}$ is the average clocking rate of the ADC. α is the activity factor of the ADC, indicating the fraction of the time that it is working at the maximum clock frequency (this depends on the input signal). $f_{\text{BW,ADC}}$ is the signal bandwidth of the ADC.

The proposed LCADC topology offers other advantages beyond the increase in power efficiency. Traditional LCADCs have an asynchronous output, which is difficult to use in applications where conventional clocked processing/transmission is required. The analysis in the previous chapter has shown that adding a retimer, a time-to-digital converter (TDC), to an asynchronous LCADC introduces an overhead in data rate, undoing much of the advantages of the LCADC. With the proposed adaptive clocking LCADC, the ADC output is always synced to the master clock; hence it is compatible with both synchronous and asynchronous data processing circuits. Moreover, if the data are directly streamed off chip, no timer is required, as the data of the ADC can fully be recovered by having a clock running

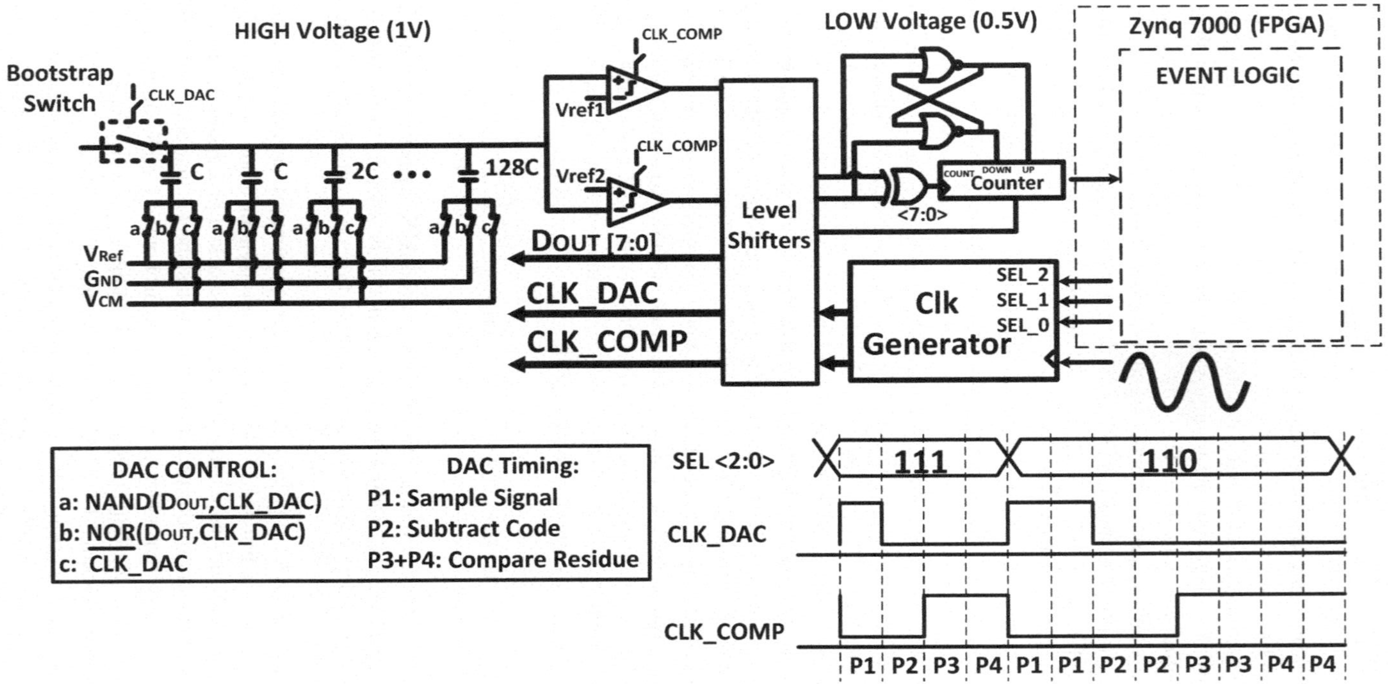

Fig. 4.2 Overview of the proposed adaptive clocking level-crossing ADC and its timing diagram. All circuitry is integrated on the prototype chip, except the event logic (which is implemented on an FPGA)

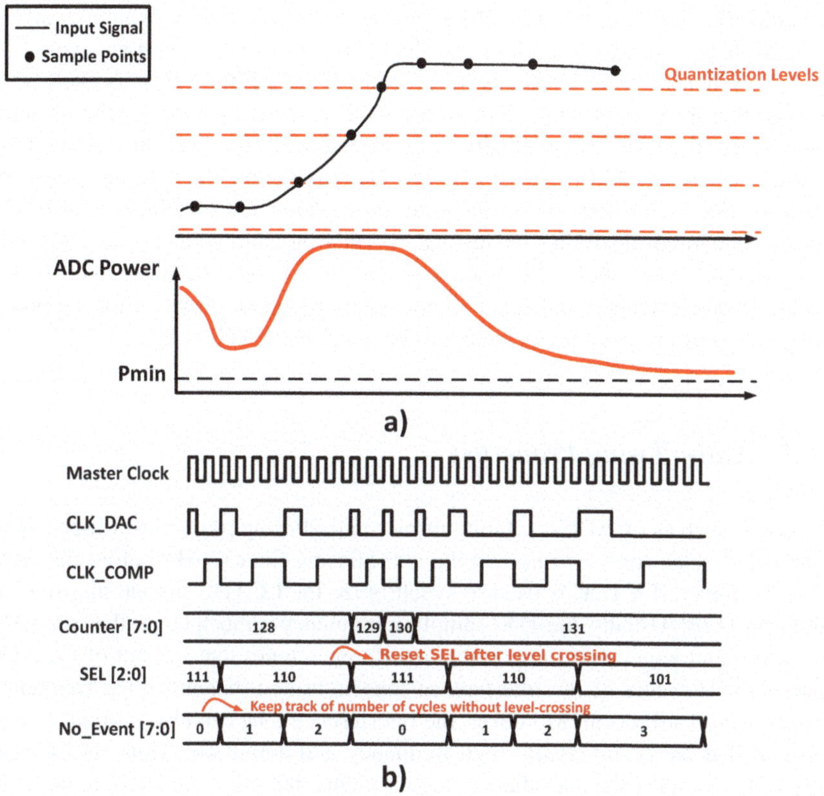

Fig. 4.3 (**a**) Clocked level-crossing sampling scheme with adaptive ADC power and (**b**) detailed timing diagram of the proposed level-crossing ADC

at the same rate as the master clock. With, for example, Manchester encoding the output data stream and the receiver clock can be synchronized [48]. Another advantage that the proposed LCADC offers is that it uses a single DAC instead of two like in traditional LCADCs (see Fig. 3.2), which leads to a significant area reduction over previous LCADCs.

4.1.2 The ADC Timing

The ADC algorithm (event logic) works as follows (see Fig. 4.3). After a new level crossing has occurred, the SEL[2:0] is set to the maximum value, and the ADC operates at maximum clock frequency. A new sample is taken by the ADC and compared to the threshold voltages. If no new level crossing occurs, the "No_Event" register counts up 1 cycle. If the No_Event register increments beyond a certain threshold (0,2,4,8,...), the SEL[2:0] signal is decreased with 1, which results in the ADC being clocked at a clock rate half the speed of the previous cycle (see Fig. 4.3). When no level crossings have occurred during the previous cycle, it is assumed that the current input slope of the ADC is lower than the maximum input slope; hence, the ADC can gradually be clocked at a slower rate. Once a new level crossing occurs, SEL[2:0] is reset again to the maximum value.[1] In our prototype design of Sect. 4.3 below, the event logic is implemented off chip on an FPGA to allow maximum flexibility for testing. As only an extra counter and a block to detect whether a threshold is crossed are required, it could be integrated on-chip at a low power overhead. In Sect. 4.3 an estimation will be provided for the power consumption of the event logic, were it to be integrated on chip.

4.1.3 Anti-aliasing Properties

CT ADCs, such as LCADCs, exhibit intrinsic anti-aliasing properties, which can be exploited to relax (or even remove) the anti-aliasing filter (AAF) before the ADC [178]. However, if a TDC is used to synchronize the LCADC output, aliasing can still occur [178]. Usually, the TDC sampling frequency is much larger than the ADC bandwidth and larger than the possible interference tones that fall out of the ADC bandwidth. However, if the interference tone is at a sufficiently high frequency, aliasing would still occur. Moreover, the interference tone can also introduce slope overload if it has a sufficiently high frequency and amplitude. Thus, an LCADC with TDC can relax the anti-aliasing requirements, but not completely remove the AAF.

The proposed topology, when operating at a fixed-rate clock, is exactly equivalent to a conventional LCADC with TDC. Thus, while it exhibits aliasing at very high interference tone frequencies or starts to slew due to slope overload, it still relaxes the AAF requirements. When using the adaptive clock, the sampling frequency is adaptively scaled down, reducing part of the anti-aliasing properties. After each sample, however, the sample clock is reset to the highest sampling rate. If a high-frequency tone is present at the input, the adaptive clock works continuously at this highest sampling rate, and hence, the degradation of the anti-aliasing properties would be minimal. To verify this, Fig. 4.4 shows a MATLAB simulation of the spectrum at the output of the proposed LCADC, in which an in-band tone at 7 kHz and an interference tone at 50 kHz are applied, with the clock rate fixed (red spectrum) and in adaptive clock mode (blue spectrum). As can be seen, no aliasing takes place in the adaptive clock mode.

[1] We also simulated another algorithm in MATLAB, in which SEL[2:0] was not reset to the maximum value after each level crossing, but gradually increased again if several level crossings occurred within a certain time window. However, this approach leads to a large degradation of the signal quality.

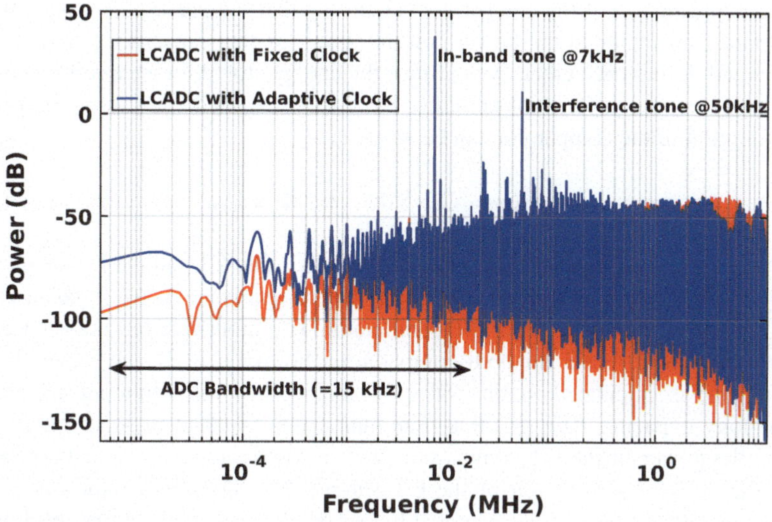

Fig. 4.4 Simulated spectrum of the adaptive-clocked LCADC, where a 50-kHz interference tone is present at the input

4.2 Comparison with the CT LCADC

To compare the power efficiency of a CT LCADC to the proposed adaptive clocking LCADC, the power consumption of a CT comparator (based on the power models derived in the previous chapter) is first compared to that of a DT (latch-based) comparator, which is used in this work to improve the power efficiency of the LCADC architecture.[2] DT comparators typically consist of two stages (often merged into one block) [10, 117, 127]: a preamplification stage and a latch stage (see Fig. 4.1). Unlike CT comparators, due to the positive feedback of the latch, only a modest gain is required to toggle a DT comparator, and due to the switched operation, the power of the comparator depends on the clock rate. For low clock rates, the DT comparator is therefore noise limited. In [10] the following equation has been derived for the integrated input-referred noise power of a double-tail latched comparator:

$$V^2_{\mathrm{n,DTCMP}} = \frac{2 \cdot \mathrm{k} \cdot \mathrm{T} \cdot n}{Av_{DT} \cdot \mathrm{C_p}}, \tag{4.3}$$

with Av_{DT} the gain of the preamplifier, n the weak inversion slope factor, and $\mathrm{C_p}$ the load capacitance of the preamplifier. This load capacitance also determines the

[2] The technology values that were used are taken from a 40 nm CMOS technology. This is the same technology used to implement the prototype circuit design in the later sections.

power consumption of the comparator (assuming that the preamplifier stage is the dominant contributor to the power consumption), as this capacitance is reset and discharged during each clock cycle. Since the energy stored for a capacitor is equal to $CV^2/2$ and the differential amplifier has two equal load capacitors, the power consumption of the comparator can be written as

$$P_{DTCMP} = C_p \cdot Vdd^2 \cdot \alpha \cdot f_{CLK,max}, \tag{4.4}$$

with $\alpha . f_{CLK,max}$ the average clock rate of the ADC. The CT and the DT topologies will now be compared to each other. For this comparison, Av_{DT} is taken equal to 10, a reasonable value for a DT comparator. The input-referred noise (IRN) is taken equal to 0.1*LSB for both comparators. From Eq. (4.3), the required load capacitance is calculated. If this value is smaller than the minimum achievable technology capacitor (0.7 fF), this minimum value is taken as load. Figure 4.5 shows the power of the DT comparator (with a worst-case $\alpha = 1$) and compares it to the power consumption of the CT variant. The DT comparator's power is determined by the noise requirement at higher resolutions, while at low resolutions the technology (minimum capacitance) limits the power consumption. Thus, for similar specifications as the CT comparator, the DT variant consumes much less power (e.g., 10× less for 8-bit resolution). Since the clock frequency of the comparator is adaptively changed in the proposed design, depending on the input signal, α is smaller than 1. Figure 4.6 shows the power consumption of the DT comparator as a function of the activity factor α, with the CT comparator's power consumption plotted as reference. Clearly, the proposed LCADC can significantly

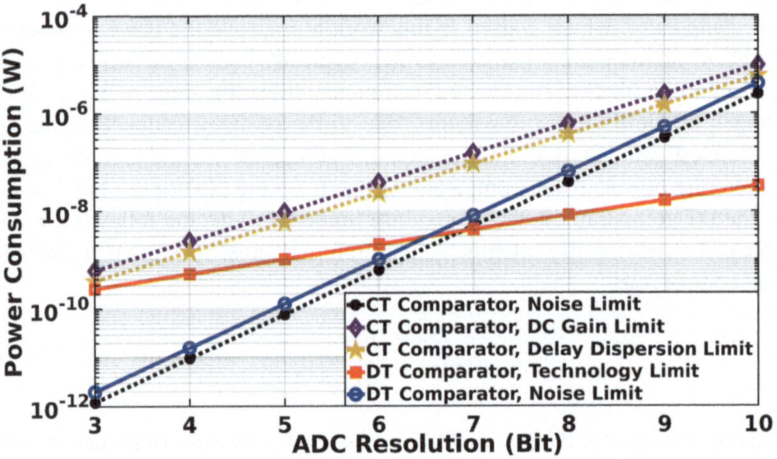

Fig. 4.5 Power consumption of a continuous-time comparator (dotted lines) and a discrete-time comparator working at a fixed clock rate (full lines), as a function of the ADC quantizer resolution. A 15-kHz sine is taken as input signal

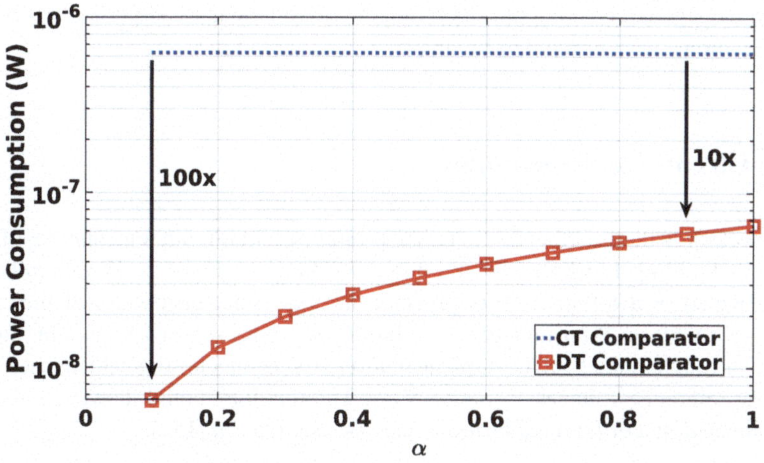

Fig. 4.6 Power consumption of a discrete-time comparator for an 8-bit ADC with 15-kHz BW, as a function of the activity factor α

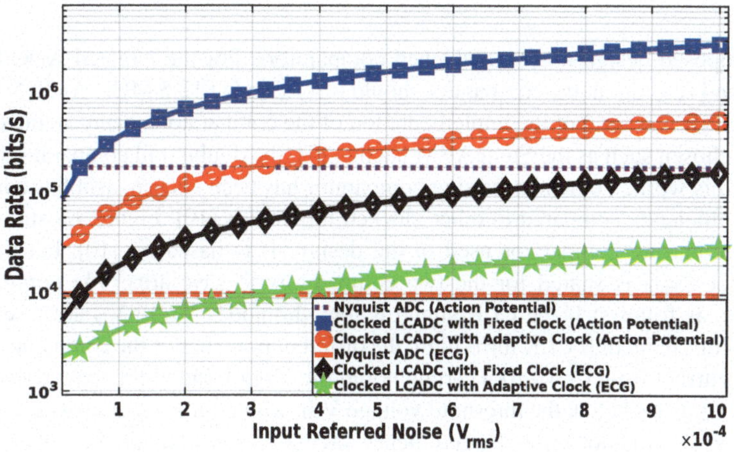

Fig. 4.7 Number of false samples with and without adaptive clock enabled, for an ECG and AP input signal

reduce the required power for the comparator for low-activity signals (almost by 100× less for $\alpha = 0.1$).

Another difference between the CT and the adaptively clocked LCADC is the generation of false level crossings due to noise. By reducing the sampling rate for slow-moving inputs, the amount of false samples taken by the ADC is decreased as well. Figure 4.7 plots the result of a MATLAB simulation with and without the adaptive clocking enabled for ECG and action potential (AP) signals (taken from [37, 40, 86]). The amount of false samples taken is significantly lower for the case with the adaptive clocking. As the adaptive clocking algorithm lags slightly behind the signal, changing the clock frequency could lead to additional errors, but in simulations these turn out to be small.

4.3 Circuit Implementation

This section will describe the detailed prototype circuit implementation of the different building blocks of the proposed LCADC of Fig. 4.2. The circuits are implemented in a 40 nm CMOS process, and the target bandwidth of the ADC is 15 kHz with an 8-bit quantizer resolution. As most sensor signals are single-ended, a single-ended ADC implementation has been chosen, which uses two external reference voltages V_{ref1} and V_{ref2}. The topology could, however, also be implemented in a differential fashion, like, for example, in [147].

4.3.1 Dynamic-Biasing Comparator

The proposed design uses two clocked comparators. For the targeted ADC bandwidth and resolution, the comparator should be clocked at 12.5 MHz. An IRN target of 0.1*LSB was set. For the implementation of the comparators, many architectures can be chosen such as the StongARM latch [117], the double-tail comparator [127], etc. In this work, a dynamic-biasing comparator has been chosen, as it has a higher power efficiency compared to other clocked topologies [10]. Figure 4.8 shows the dynamic-biasing comparator used in the design. It is based on [10], but the tail capacitor C_{hold} is shared for the two comparators to save area. The comparator operates as follows: If the CLK signal is low, the nodes V_A to V_D are reset, and C_{hold} is discharged. If CLK turns high, the different branches of the comparator start to pull current from C_{hold}. At a certain point, the Vsg of one of the input transistors (M3–M6) drops below the threshold voltage Vth, which turns the transistor OFF. In this way, C_{hold} is not fully charged, hence saving energy compared to other latch-based topologies, that fully discharge the preamplifier at each comparison [10]. The cumulative probability density of the comparator's output is plotted in Fig. 4.9, as well as a fitted normal distribution. From post-layout simulations, it was determined that the IRN value σ is 343.45 μVrms, which is sufficient for the application. The comparator's device sizes are given in the table in Fig. 4.8.

4.3.2 Event Logic and Clock Generator

The proposed LCADC uses an adaptive sample clock to save power for time-sparse input signals. The activity tracking algorithm is implemented on a Zynq 7000-based FPGA (ZedBoard) (see the overall block diagram of Fig. 4.2), while

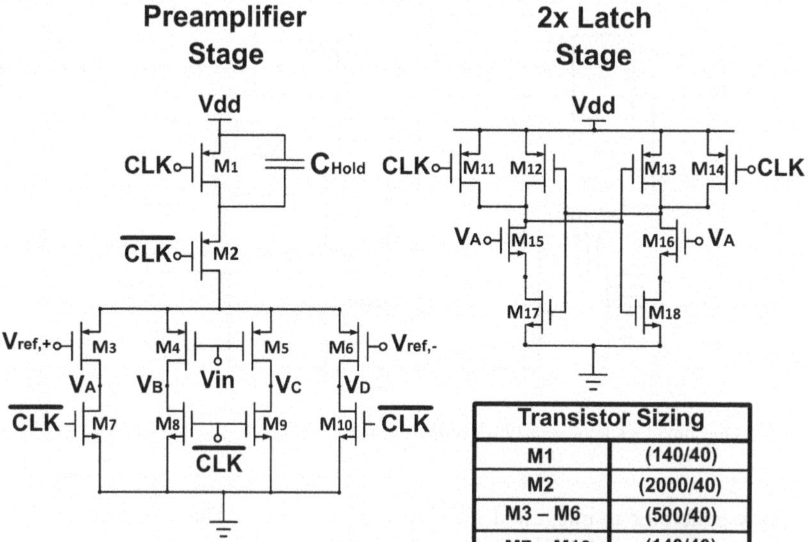

Transistor Sizing	
M1	(140/40)
M2	(2000/40)
M3 – M6	(500/40)
M7 – M10	(140/40)
M11 – M14	(280/40)
M15 – M18	(140/40)
C_{Hold}	5.76 fF

Fig. 4.8 Dynamic-biasing comparator with shared hold capacitor and table with the different device sizes

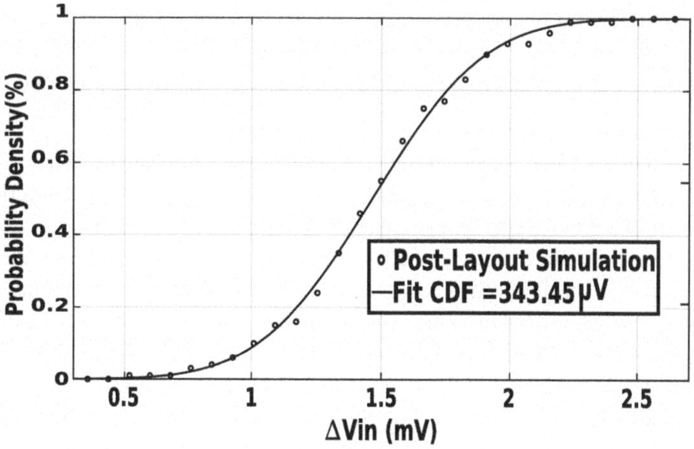

Fig. 4.9 Cumulative probability density of the dynamic biased comparator's output, calculated from post-layout simulations

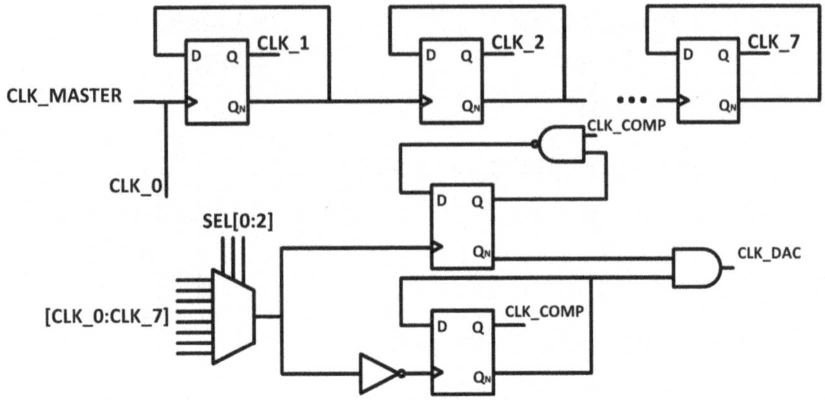

Fig. 4.10 On-chip clock generator block of the LCADC

the clock generator is integrated on chip (see Fig. 4.10). The input signal of this clock generator is a 25-MHz clock (CLK_MASTER), which is down-converted by a divide-by-2 clock divider, made up of a chain of D flip-flops. The output clocks of these flip-flops are passed on to a multiplexer block, which is controlled by the external SEL signal, coming from the FPGA. SEL chooses which clock signal is used, and from this signal, two clock signals with a duty cycle of T/4 (for CLK_DAC) and T/2 (for CLK_COMP) are generated. Via the level shifters, these signals are passed to the analog circuits. A full timing diagram of the algorithm has been given in Fig. 4.3. The event logic has been implemented on FPGA because of the flexibility to be able to experiment with different types of algorithms. Eventually, a lightweight algorithm has been used, consisting only of an extra counter and threshold detectors. To assess the power overhead of this event logic if it were to be integrated on chip, an estimation of the power consumption is provided below.

Estimation of the Event Logic's Power Consumption

As discussed in Sect. 4.1, the event logic monitors each sample cycle if a new sample is taken. If no sample is taken, the counter that keeps track of the amount of cycles without a threshold crossing is increased. Once this counter reaches a multiple of 2, the event logic updates the SEL[2:0] signal, which scales down the ADC clock frequency. This event logic thus requires an extra counter, logic gates that detect if the counter reaches a certain value, and a register that stores the SEL[2:0] signal. An 8-bit counter is sufficient, as the maximum value it should store is 255. The SEL [2:0] signal needs to toggle at most eight times. Hence, logic gates to detect eight counter values need to be integrated. Assuming a (naive) implementation using two-input NAND gates, in which each bit of the counter is compared via these NAND gates to the values at which SEL [2:0] should change, 15 gates are required per counter value. Hence, a total of 120 gates are required. The SEL[2:0] signal can be stored in a 3-bit register. However, since this register is only updated sporadically, it is neglected in this power estimation. Assuming that a ripple counter is used (like the one already implemented in the ADC), the power of the 8-bit counter $P_{counter}$ can be estimated as

$$P_{counter} = \sum_{n=1}^{8} \frac{C_{D\text{-}FF} \cdot Vdd^2 \cdot \alpha \cdot f_{CLK,max}}{2^{n-1}}, \tag{4.5}$$

where $C_{D\text{-}FF}$ is the total gate capacitance of the D flip-flop used in the counter, Vdd is the supply of the digital logic (0.5 V in the presented design), $f_{CLK,max}$ is the maximum clock frequency at which the ADC is clocked (12.5 MHz in the prototype IC, half of the input clock), and α is the activity factor which indicates the fraction of the time the ADC is working at the maximum clock frequency.

The power of the integrated NAND gates P_{NAND} is estimated by

$$P_{NAND} = 120 \cdot C_{NAND} \cdot Vdd^2 \cdot \alpha \cdot f_{CLK,max} \tag{4.6}$$

with C_{NAND} the gate capacitance of a NAND gate. Figure 4.11 plots the estimated power of the event logic as a function of the activity factor, which depends on the signal. Assuming that $\alpha = 1$ for a 15-kHz sine input (which is an overestimation), the event logic would consume 270 nW. The additional power of the event logic thus only slightly degrades the ADC power efficiency. Moreover, the event logic can work at the variable ADC clock rate, making that the power consumption of this logic scales with the signal activity.

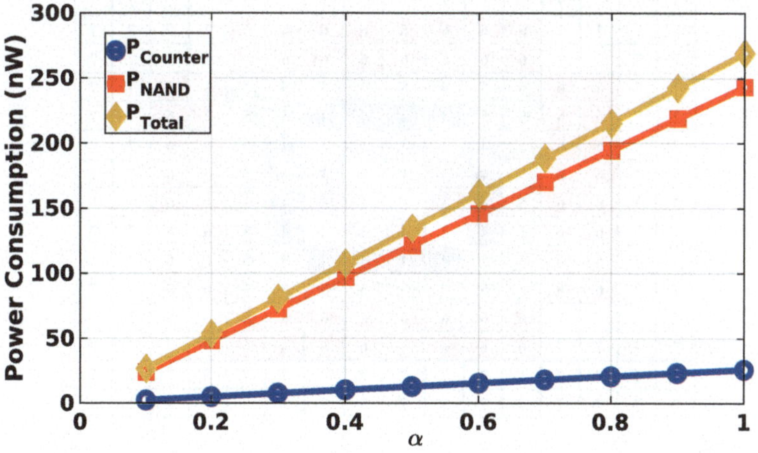

Fig. 4.11 Power consumption of the event logic as a function of the activity factor α

4.3.3 ADC Logic and Counter

If a level crossing is detected by the comparators, the counter of the ADC needs to be updated. This is done by means of an SR-latch that determines the direction of the level crossing (up or down) and a modified ripple counter (see Fig. 4.2). Depending on which comparator toggles, the SR-latch is reset or set, which controls the counter direction. The output of the comparators is connected to an XOR gate, which is given to the counter as a clock. If a level crossing is registered, the counter is updated, and this value is sent to the FPGA.

4.3.4 CDAC and Bootstrap Switch

For the proposed LCADC, only a single DAC is needed, unlike previous topologies that needed two DACs [171]. Because of its simplicity, a binary-weighted DAC has been implemented (see Fig. 4.2), with a unit capacitance of 2.4 fF. This capacitor size has been chosen to make sure that the capacitive DAC (CDAC) noise and mismatch do not limit the design specifications. The CDAC layout is placed in a common-centroid configuration, as depicted in Fig. 4.12. At each ADC cycle, the input signal is sampled on the CDAC with a bootstrap switch (see Fig. 4.13), which has been designed as described in [116].

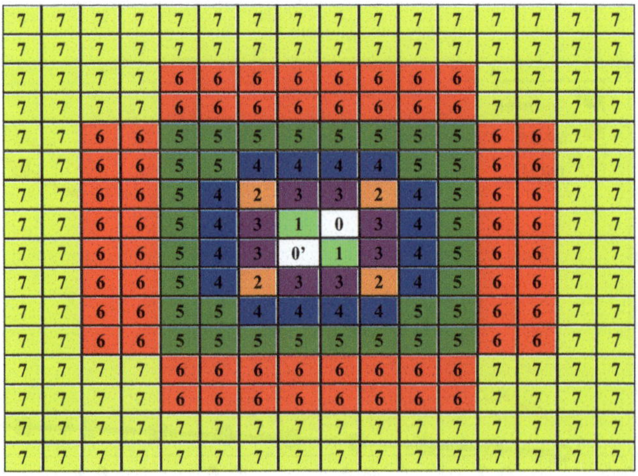

Fig. 4.12 CDAC layout scheme, in common-centroid configuration. Dummy capacitors placed around the CDAC are not depicted

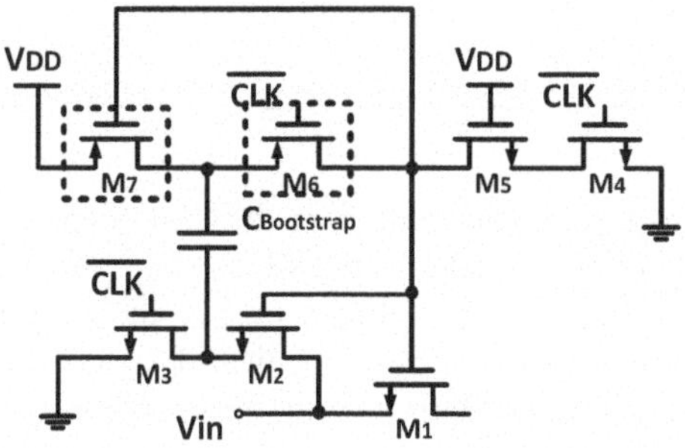

Transistor Sizing	
M1	(480/80)
M2	(480/80)
M3	(140/40)
M4	(140/40)
M5	(500/40)
M6	(140/40)
M7	(140/40)
$C_{Bootstrap}$	100 fF

Fig. 4.13 Bootstrap switch used in the design and its corresponding device sizes

Design Tip: CDAC Design and Layout Considerations

The CDAC is used in many data converters like the SAR ADC and also in LCADCs. Its power consumption is proportional to the unit capacitors used in the DAC; hence, using small capacitors allows to reduce the DAC power. Often metal-oxide-metal (MOM) capacitors (MOMCAPS) are used, which are made in the metal layers of the CMOS process, due to their excellent linearity and high capacitor density. The capacitors in the CDAC can be technology, noise, or mismatch limited:

- The minimal unit capacitance of MOMCAPS in older CMOS technologies was limited by the limited lithography resolution, which determines the minimum spacing and capacitor area; hence, only relatively high unit capacitor values could reliably be fabricated. In modern CMOS technologies, however, scaling down the capacitors is not an issue anymore, with values down to 100 aF being possible [44]!
- The sampling operation in, e.g., a SAR ADC or the proposed LCADC introduces kTC-noise. If this noise limits the SNR of the ADC, the ADC is noise limited, and sizing up the capacitors seems the only choice for a designer (even though some tricks like kTC-noise cancelation are possible [70]).
- For most medium-resolution designs (like the LCADCs that are considered in this book), mismatch is likely a bigger issue than noise. Certainly in LCADCs, which are mainly limited by harmonic distortion, as discussed in the previous chapter, the mismatch of the CDAC should be taken into account by the designer (even though it is not reflected too much in the available literature on LCADCs).

(continued)

Mismatch in capacitors can mainly be introduced because of three sources: random mismatch, process-induced mismatch, and designer-induced mismatch.

- Random mismatch is caused by random fluctuations in the CMOS process and follows the well-known Pelgrom's model [106], which states that the mismatch of components decreases with the square root of the area of the component. To tackle random mismatch in a practical design, either the size of the capacitors can be chosen larger or the mismatch can be compensated by means of a calibration technique.
- Process-induced mismatch is, as the name suggests, mismatch that originates not from random fluctuations but is introduced more systematically in the CMOS process. This can be due to machine imperfections, etching effects, etc. In the CDAC, this can lead to gradient errors, which makes that the capacitors can have linear, cubic, or even higher-order spatially correlated mismatch. Typically these gradients are not well modeled in a PDK, so a designer needs to take precautions to avoid these effects. For example, the CDAC layout often is placed in common-centroid configuration, as is done for the DAC in the design from this chapter (see Fig. 4.12). This common-centroid approach can be pushed quite far, with complex layout schemes that distribute the unit capacitors from each bit of the DAC in a "chessboard" pattern [68].
- The last type of mismatch is the designer-induced mismatch. An inexperienced designer might look at literature and initially think that a complex layout pattern is the best layout scheme for the ADC, as this would eliminate any type of gradient effect in the array. However, the routing of such array can become very complex, resulting in long (and overlapping) routing lines that introduce parasitic capacitors [68]. These routing parasitics introduce many issues in the CDAC. For example, the ADC operation can severely be degraded, as the parasitics can increase the sample capacitance for the input switch (which then needs to be dimensioned for a larger load); larger routing parasitics can also cause settling issues for the DAC driver due to the increased load capacitance of the routing, etc. Moreover, the routing can increase the ADC power consumption; in some designs the power consumption doubles due to layout parasitics [45]. Lastly, in the worst case, the routing introduces mismatch itself, if, for example, the routing is not symmetrical [68, 148].

Hence, "less is more" also holds here. A simpler layout like [46] can be preferable over a complex layout scheme that minimizes the effect of gradients. Certainly for medium- or low-resolution ADCs, the gradients might not be the largest error in the CDAC. Another good practice is to use MOMCAPS in higher metal layers (which still have the minimum metal

(continued)

width), to reduce the parasitics to the substrate. Moreover, sufficient dummy capacitors should be placed (in every direction) around all the capacitors that are used in the CDAC. Lastly, to avoid surprises due to "designer faults" in the layout, it is also important to simulate the CDAC as soon as possible at the post-layout level!

4.3.5 Level Shifters

To reduce the power consumption of digital circuits of the ADC, the counter and clock generator circuits operate at a reduced supply voltage of 0.5 V. However, to maximize the input swing of the ADC, the analog circuits require a full-scale input of 1 Vpp and hence work at a supply of 1 V. Since the control signals need to cross different voltage domains, level shifter circuits are used to pass the signals from one domain to the other. These circuits are depicted in Fig. 4.14.

4.4 Measurement Results

A prototype chip with 8-bit LCADC and 15-kHz signal bandwidth (SBW) has been fabricated in a 40 nm CMOS technology. Figure 4.15 shows the die photo and the prototype chip that is wire-bonded to a small measurement PCB. The total chip area is $1.5 \times 1.5\ mm^2$, with $0.012\ mm^2$ of active area.

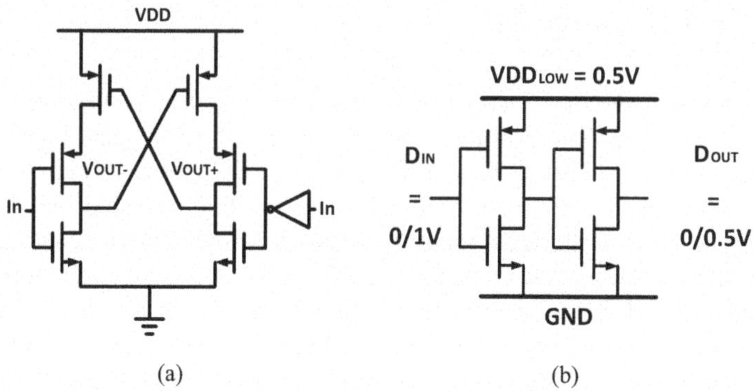

Fig. 4.14 Level shifter circuits, from (**a**) the 0.5 V to the 1 V domain [150, 189] and (**b**) from the 1 V to the 0.5 V domain

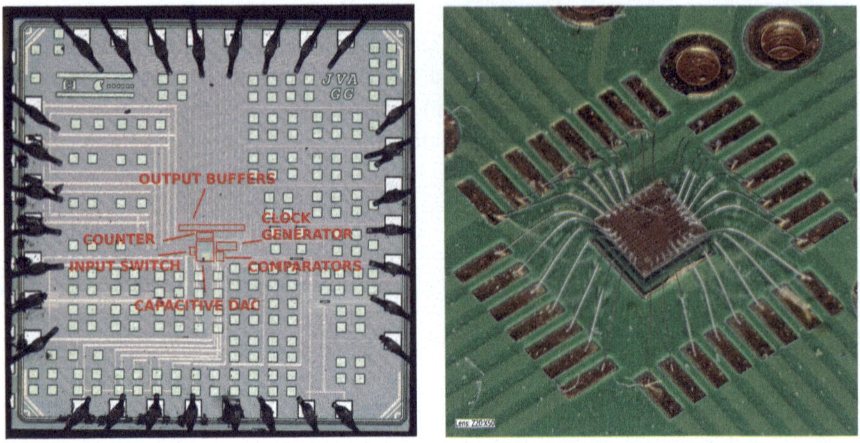

Fig. 4.15 Die photograph (left) and chip bonded to the PCB (right)

4.4.1 *Measurement Setup*

To measure the prototype chip, a fully automated measurement setup has been used, as shown in Figs. 4.16 and 4.17. The ADC master clock of 25 MHz is provided externally by a pulse pattern generator; the input signal is provided by an arbitrary waveform generator (AWG). The ADC data are sent to a Zynq7000 SoC (Zedboard) on which the adaptive clocking algorithm is implemented, as well as the logic to record the ADC output. The ADC reference voltages are provided via external source meter units.

4.4.2 *Chip Measurements*

Figure 4.18 shows the measured spectrum for a sine input of 10 kHz, resulting in a signal-to-noise-and-distortion ratio (SNDR) in the 15-kHz signal band of 62.5 dB. Figure 4.19 shows the in-band SNDR of the ADC (for a 10-kHz tone) as a function of the input amplitude. Figure 4.20 shows the measured SNDR in the 15-kHz signal band for different input frequencies, before and after calibrating the mismatch in the DAC and the comparator offset by means of a one-time foreground calibration with a 10-kHz sine wave input. For this calibration step, an offline script calculates the best polynomial fit to the 10-kHz sine wave; this best fit is then applied to each ADC recording. Since the signal bandwidth is much smaller than the sampling rate (similarly to an oversampling data converter), the in-band quantization noise is lower compared to that of a classical 8-bit Nyquist ADC (as discussed in Sect. 3.2.1); hence, a higher SNDR is reached. The peak SNDR of 10.4 ENOBs (64.4 dB) is reached at 15 kHz. At higher frequencies, the ADC experiences slope overload (the clocked ADC cannot sufficiently track the input signal anymore),

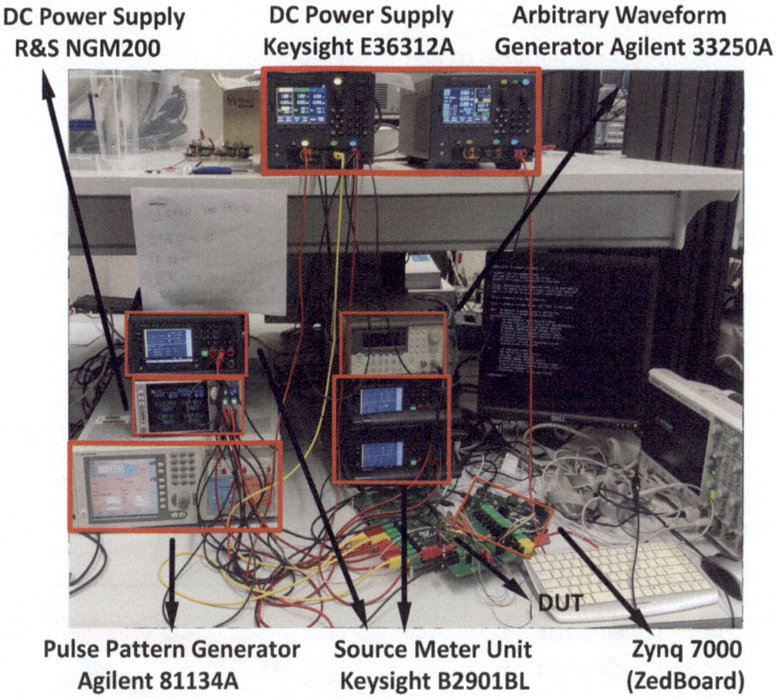

Fig. 4.16 Overview of the measurement setup

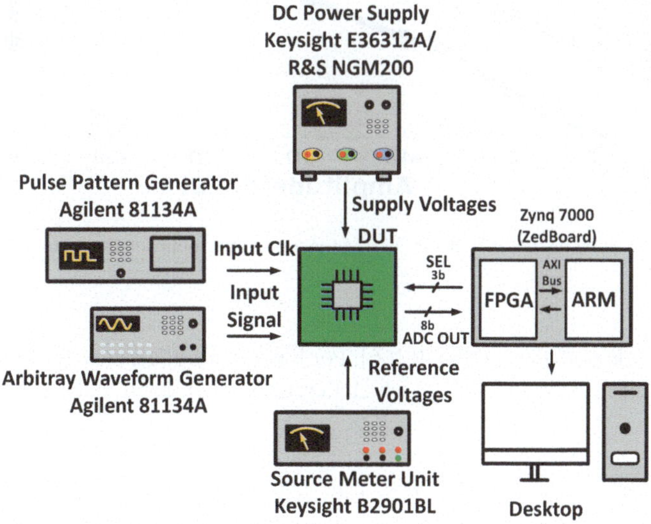

Fig. 4.17 Block diagram of the measurement setup

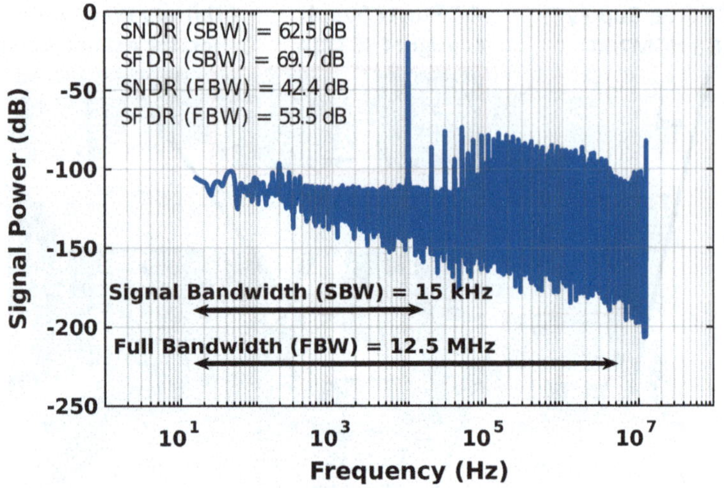

Fig. 4.18 Output spectrum for a 10-kHz input sine (15M-point FFT)

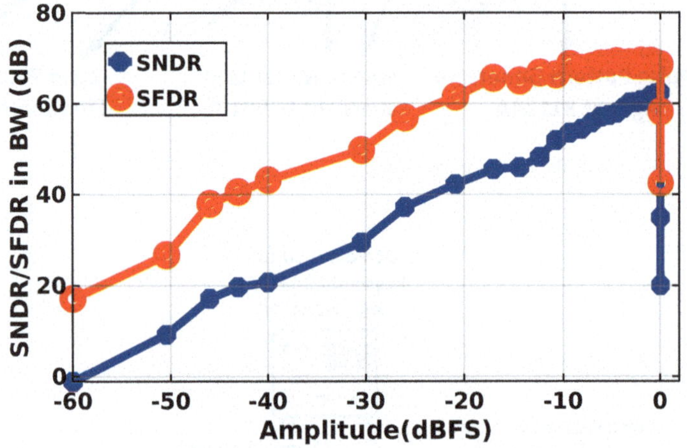

Fig. 4.19 Measured in-band SNDR/SFDR as a function of the input amplitude, for a 10-kHz input signal

leading to a degradation in SNDR [66]. For low input frequencies, the harmonics of the ADC fall in-band [62, 177],[3] which degrades the in-band SNDR. As described

[3] Some LCADC designs like [66, 148] also have an SNDR degradation at lower frequencies due to leakage effects, which would be more difficult to calibrate. Since the proposed ADC works in a clock-based manner, the DAC and comparator are reset each cycle; thus, as long as the designed CDAC can hold the charge for the minimum clock period, leakage effects will not degrade the SNDR.

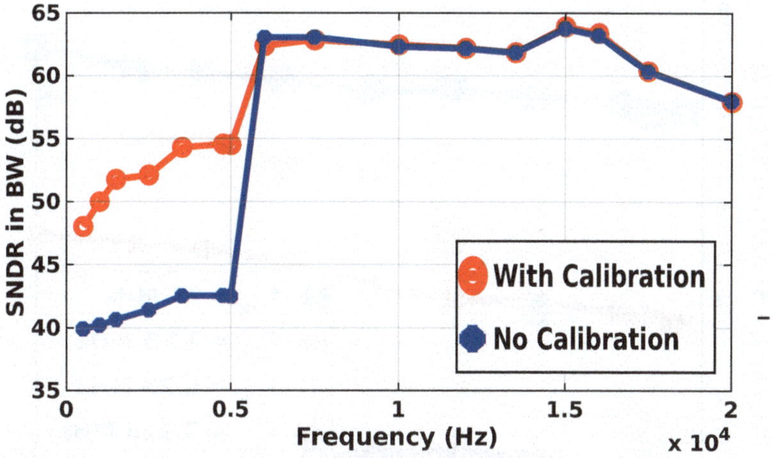

Fig. 4.20 Measured SNDR in the 15-kHz signal band for different input frequencies, before and after one-time foreground mismatch calibration

in Sect. 3.2.2, the mismatch of the DAC and comparator contributes a lot to the harmonics. After calibration, a 10-dB increase in SNDR is noted at low input frequencies.

Figure 4.21 shows the power consumption versus the input sine frequency, for different fixed clock frequencies (in this mode, the SEL control signal is kept constant), as well as in event-driven mode (when the event logic adapts the ADC clock frequency based on the signal activity). In the fixed mode, once a clock frequency is chosen (as a function of the slew rate of the input signal), the power consumption varies only weakly with the input frequency. As the ADC only generates a few discrete clock frequencies, it works inefficiently in the fixed mode. In contrast, in the event-driven mode, the power consumption varies from 0.92 μW for an ECG input signal (see Fig. 4.23) to 5.38 μW for a 15-kHz full-scale sine wave (excluding I/O and clock buffers). This highlights the activity-based power consumption of the ADC. In Fig. 4.22, the ADC power breakdown is shown for two input frequencies. It can be noted that at lower input frequencies, the DAC and comparator clock rate are reduced, and most power is consumed by the digital logic, which is always running (e.g., the clock generator circuit) or leakage (e.g., the DAC driver logic and the counter logic). These ultimately set a lower bound on the power consumption that the ADC can achieve. For a high input frequency, the ADC power scales with the higher sampling frequency, and the DAC and comparator consume more power, as indicated in the power breakdown.

In the previous chapter, the system-level advantage of the LCADC (mainly reduced transmission/processing power) has been highlighted. To illustrate the power savings in any subsequent digital blocks, Fig. 4.23 plots the power of the I/O buffers (i.e., the logic gates that drive the data off chip), showing a reduction of 3× for an ECG signal. Clearly, the I/O buffers consume significantly less, thanks to

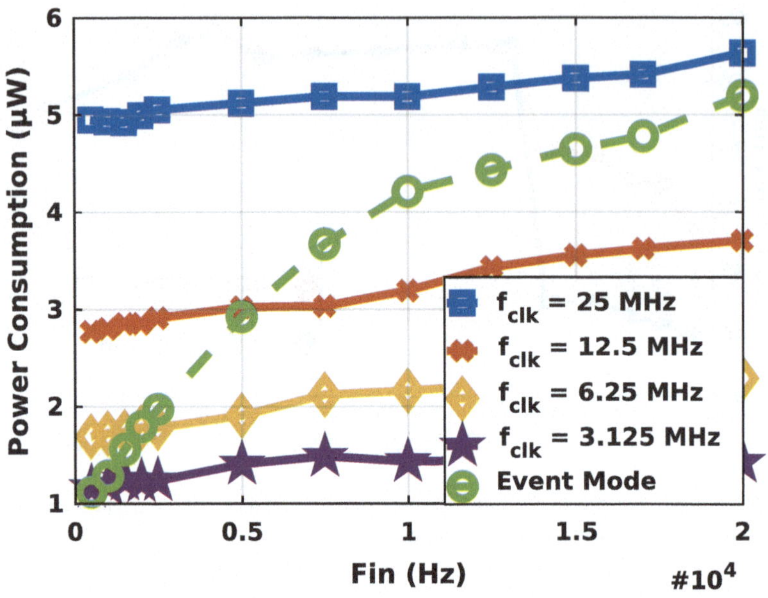

Fig. 4.21 Measured power consumption as a function of the sine wave input frequency in fixed and event-driven modes

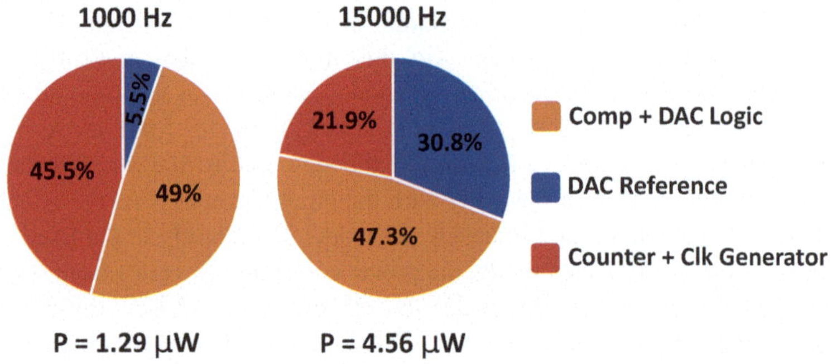

Fig. 4.22 Power breakdown of the ADC for a 1-kHz and 15-kHz input signal

the signal-dependent output of the LCADC, which achieves a 30% reduction in data compared to a Nyquist-rate ADC for this ECG signal.

4.4.3 Comparison with the State of the Art

Table 4.1 shows a comparison of our prototype design with the achieved performances of recently published LCADCs. Our design achieves an ENOB of 10.4 bits,

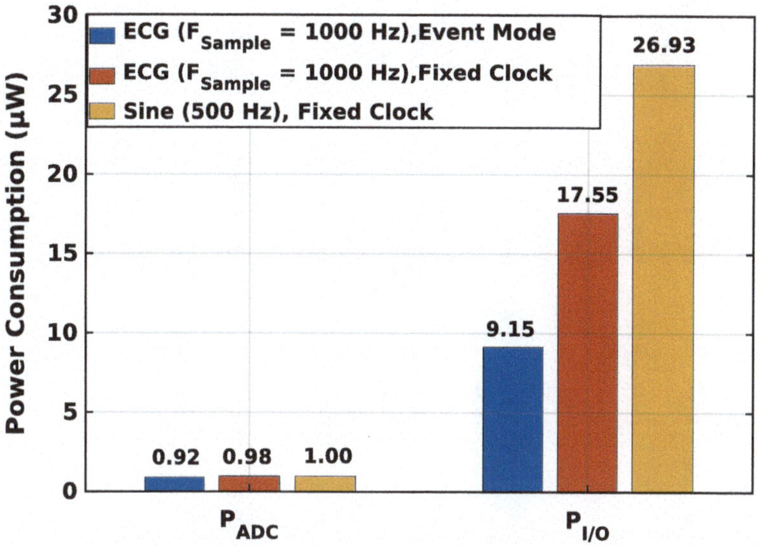

Fig. 4.23 Comparison of the power consumption of the ADC and the I/O buffers, for an ECG signal and a 500-Hz sine wave as input signals

which is among the highest reported in literature for LCADCs. The table also lists various FoM values that the prototype chip achieves. They are described next.

4.4.3.1 DR Figure of Merit

For LCADCs used for sensing applications, the dynamic range (DR) is the most important accuracy metric, and hence, the (Schreier and Walden) DR FoM is plotted in Fig. 4.24.[4] Given are both the best-case operating scenario (when the ADC has an input signal with low activity such as, e.g., ECG) and the worst-case scenario (when a high-frequency input is applied, like a sine wave near the Nyquist frequency). The proposed prototype is among the best designs in literature, reaching a best-case Walden DR FoM of 119.8 fJ/conv.step and a best-case Schreier DR FoM of 152 dB. For higher DR designs (designs with >6 bit DR), it achieves the best FoM of all published designs (in the best-case scenario, for which the ADC is designed). Even if the event logic would be integrated on chip (which would increase the power in the worst-case scenario with 270 nW), the ADC would still be among the best LCADC designs in literature, validating the proposed adaptive clocking approach.

[4] The textbox below discusses in more detail why this FoM is used.

Table 4.1 Comparison with state-of-the-art level-crossing ADCs

	Weltin-Wu	Wang	Wu	Timmermans	Wang	Patil	Li	
	JSSC13 [171]	JSSC20 [164]	JSSC17 [178]	VLSI23 [147]	JSSC21 [168]	JSSC16 [103]	ESSCIRC23 [64]	This work
Application	Sensors	IOT	RF	EMG	Wake-up circuit	RF	ECG	Biomedical sensing
Topology	AR LCS	AR LCS	LCS	LCS	LCS	Integrated LCS	SAR+LCS	LCS
Technology (nm)	130	28	65	65	180	28 FDSOI	55	40
Supply voltage (V)	0.8	/	1	1.2	0.6	0.65	1.2	0.5/1
ADC resolution	8b	7b	4b	6b	5b	/	4b+4b	8b
Discrete DSP compatible	No	Yes	Yes	No	Yes	No	No	Yes
Area (mm^2)	0.36	0.0126	0.3	0.005	/	0.0032	0.063	0.012
Power consumption (μW)	3–8.5	146–205	30000	0.076–9.9	0.075–1	24	0.44–7	0.92–5.38
Bandwidth	20 kHz	1.42 MHz	19 MHz	16 kHz	80 kHz	40 MHz	25 kHz	15 kHz
ENOB (bits)	7.5–8.7	6.87–8.6	9.7	9.9	/	5–6.7	8.2–10	10.4
SNDR (dB)	47–54	43.12–53.53	59.9	61.6–72	/	32–42	51.2–62.2	48–64.4
SFDR (dB)	58.3 (1 kHz)	58	56	/	/	34	/	81.8
Walden SNDR FoM (fJ/conv)*	210–880	186.1–439.4	977	3–95	/	3–10	29.7–171	138–149.4
Walden DR FoM (fJ/conv)**	293–830	401.6–564	49342	37.1–4834	13.9–205.1	/	550–8750	119.8–700.5
Schreier SNDR FoM (SNDR) (dB)***	147.7–145.2	152–143	147.9	174.8–164	/	164.2–154.2	158.74–155.5	158.85–150
Schreier DR FoM (dB)****	148.16–143.5	143.8–142.3	113.9	151–130	152–140	/	133.38–121.37	152–144.37

*FoM $= \text{Power}/2 * f_{Nyquist} * 2^{ENOB}$

**FoM $= \text{Power}/2 * f_{Nyquist} * 2^{DR}$; DR is taken equal to the quantizer resolution

***FoM $= \text{SNDR} + 10.\log(\frac{f_{Nyquist}}{\text{Power}})$

****FoM $= \text{DR} + 10.\log(\frac{f_{Nyquist}}{\text{Power}})$; DR is taken equal to the quantizer resolution

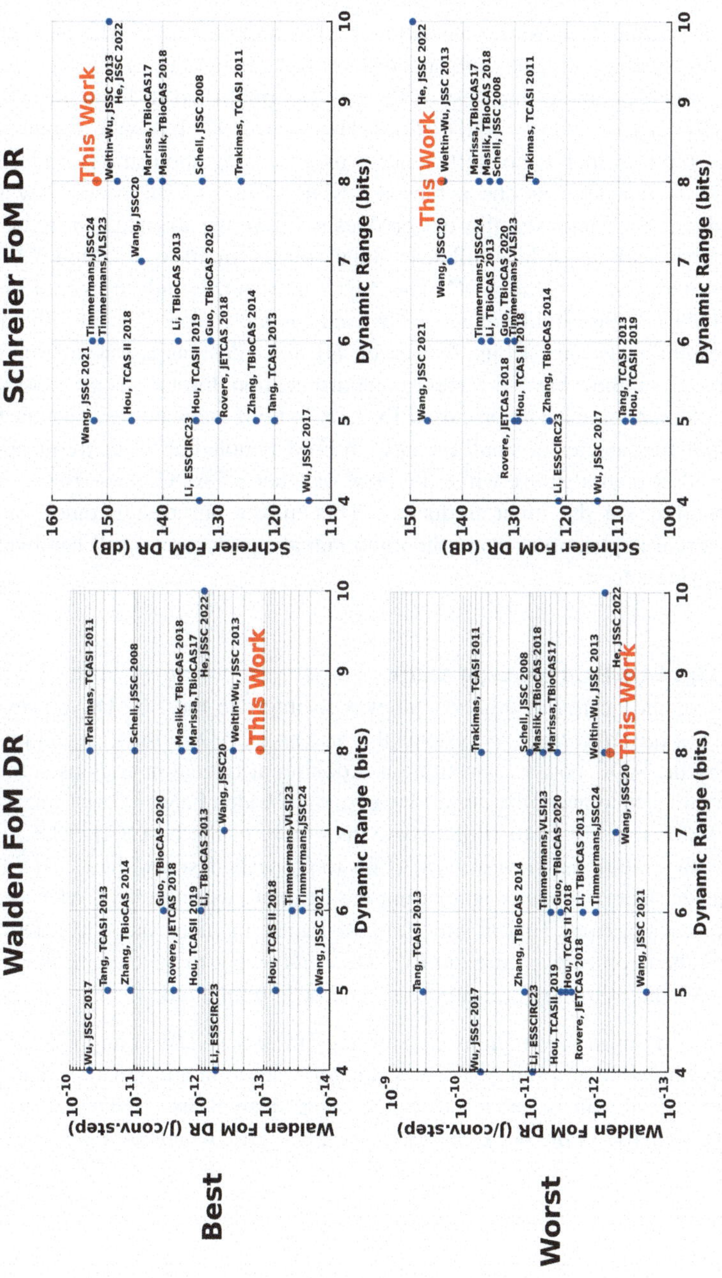

Fig. 4.24 Overview of state-of-the-art LCADC implementations from literature, in the best-case (low input activity) and the worst-case (high input activity) scenario. To calculate the FoMs, the DR is used as the metric for the ADC accuracy

4.4.3.2 SNDR Figure of Merit

Traditionally, the SNDR is also used as an accuracy metric for LCADCs, since this metric is commonly used in conventional ADC architectures.[5] The presented prototype chip reaches a peak Walden FoM of 138 fJ/conv. (at a 15-kHz input frequency), which is among the best of the designs published in literature. Only [64, 103, 147] achieve a higher SNDR FoM. However, [147] has only a quantizer resolution of 6 bits, which is not sufficient for most sensing applications. In [103], a good FoM is reached, but the achieved SNDR of 32–42 dB is too low for sensing applications. Moreover, this design does not have the advantage of reduced data throughput and has a low full-scale input range. In [64], a 4-bit LCADC is combined with a 4-bit SAR ADC. The ADC, however, is only triggered once the 4-bit LCADC's threshold voltage is crossed, which makes that the effective quantizer resolution is only 4 bits. Moreover, all these designs are asynchronous [64, 103, 147], so they have the same synchronization limitations as classical LCADC implementations. The proposed LCADC's FoM stays pretty constant for different input frequencies (a small increase in FoM is noted at lower frequencies due to the SNDR degradation), while the FoM of other LCADC topologies varies widely depending on the input frequency. This makes our design much more efficient in a real application, where the input signal can occupy a full bandwidth rather than a single tone.

Design Tip: LCADC Figures of Merit

The goal of the design in this chapter was to improve the LCADC power efficiency *at the ADC level*, as systems that use an LCADC [48, 132] can be limited by the ADC power consumption. Looking at typical metrics used to describe the ADC power efficiency, like, e.g., the Walden FoM, LCADCs can achieve FoMs that are up to 100× worse than those of the best SAR ADCs. However, the traditional FoM of ADCs fails to properly describe the LCADC performance. Hence, directly comparing the LCADC with the SAR ADC is not a fair comparison. This is why Table 4.1 only includes other LCADC designs. When comparing different LCADCs and trying to benchmark them with a FoM, the following precautions should be kept in mind:

1. As pointed out in the previous chapter, the DR and SNDR of an LCADC are decoupled, e.g., even at a low quantizer resolution and thus low DR, a high SNDR can be achieved. Since some designs use SNDR as the accuracy metric in the FoM, designs with a low DR (which are not useful

(continued)

[5] Using SNDR for LCADCs is anyway not a fair/accurate metric to describe the accuracy; see the discussion in the previous chapter on DR versus SNR decoupling and the discussion in the textbox below.

for every/any application) might be favored, as increasing the DR and thus the quantizer resolution increases the BW and/or gain requirements of the LCADC building blocks.

2. Designs from literature use often spline or polynomial interpolation to reconstruct the input signal from the nonuniform LCADC output. However, due to the large latency/compute power required for such interpolation schemes, these methods are usually not applied in real-time edge applications. Moreover, the interpolation scheme is not representative of the circuit performance. Hence, ideally, all LCADCs' performances are compared using the same interpolation method. The easiest and most fair method to use would be ZOH.
3. The input signals used to report the ADC power efficiency vary also for different papers. Some papers try to calculate the FoM via the classical sine input (such as this work); others like [147] use the power efficiency obtained when applying an actual signal such as ECG.
4. The SNDR and power consumption of the LCADC vary with the input frequency of the signal that is applied. Depending on at which point the measurement is done, the LCADC power efficiency varies significantly (up to a factor 100× [147]), which makes it difficult to compare performances. Moreover, the LCADC FoM is usually always calculated using the Nyquist frequency as the speed metric, even though the best power consumption numbers are obtained when the input frequency is much lower than this frequency. This raises the question if the current way of calculating the FoMs for LCADCs is correct.
5. Finally, the goal of using an LCADC in a system is almost never to improve the raw ADC performance (this is also not needed, and we can go pretty low in FoM already with standard ADC topologies), but to improve the system performance. However, benchmarking at the system level is difficult from only the ADC design. For example, the compression ratio (CR) the LCADC can achieve for a signal could be an indication of the system-level performance, but this CR is signal specific and is not standardized. Another option is to monitor the I/O power of the ADC (as is done in this book). To use these metrics, a common definition for CR should be agreed upon, with a fixed dataset/methodology that every designer can use to benchmark the circuit.

With all these remarks, one may think that comparing LCADC architectures is impossible, and even that, if one would try, the FoM might vary significantly depending on which one you choose (as can be seen in Table 4.1, which reports the various FoMs that exist for ADCs). Therefore, to compare LCADCs, the following suggestions are made:

(continued)

- For the accuracy of the FoM, use the metric (either SNDR or DR) which applies best to the application at hand. Often, the SNDR is reported, but for most applications in the biomedical/sensor field actually the DR would be a more important metric.
- The idea of LCADCs is that for low-activity inputs, the ADC power is reduced; hence, the minimum achievable power P_{min} is another important parameter. To give a complete view of the circuit performance, a second measurement should be done with a high-activity input, which then reports the maximum circuit power P_{max}.
- The CR or some related metrics are good to mention in the comparison table but should be omitted from the FoM, as the CR is dependent on the signal and is difficult to compare system to system.
- Therefore, to compare LCADCs at ADC level, the best FoM to use is the Schreier DR FoM, see Eq. (2.32), or a modified Walden FoM that uses DR instead of SNDR:

$$\mathrm{FoM}_{\mathrm{Walden,DR}} = \frac{P_{min}}{2^{DR} \cdot 2 \cdot f_{Nyquist}}, \tag{4.7}$$

with DR the dynamic range expressed in bits.

4.5 Conclusions

This chapter has presented the design of an event-driven level-crossing ADC (LCADC) with adaptive clocking. To reduce the large power consumption of the continuous-time (CT) comparators, an LCADC topology has been proposed in which the CT comparators are replaced by more power-efficient clocked discrete-time comparators. To keep the advantage of signal-dependent power consumption for time-sparse input signals, an adaptive clocking strategy is used. The adaptive clock is selected by an algorithm that checks the signal activity and, when possible, selects a lower sampling frequency generated from an on-chip clock generator. This adaptive clocking also helps to suppress false level crossings. A prototype LCADC chip has been fabricated in 40 nm CMOS and achieves an ENOB of 10.4 bits, a best-case Walden SNDR/DR FoM of 138/119.8 fJ/conv.step, and a best-case Schreier SNDR/DR FoM of 158.85/152 dB. The system-level benefits of the topology have also been illustrated: A 30% data reduction for ECG applications has been noted, which results in a large reduction in system power consumption. The presented LCADC design can be used in low-power edge applications, both regularly clocked and event-driven (e.g., spiking neural networks), where low-power readout, processing, and data transmission are required. Moreover, the presented

ADC opens the path to applying event-driven techniques in applications that inherently need to be synchronized, such as, e.g., multichannel sensor systems. The presented LCADC will be extended in the next chapter to a multichannel architecture and will be integrated in a system with an on-chip spiking neural network processor.

Design Errata/Improvements

As is likely the case with any academic integrated circuit design, the prototype chip presented in this chapter also suffered from some design flaws and some blocks could better be redesigned. As a guide for future designers who would like to use the presented LCADC topology and who wish not to repeat these mistakes, these errata are listed below:

- The clock generator block of the LCADC generates several clocks that can be applied to the LCADC. In the event-driven mode, the ADC switches between the different clock frequencies, which requires that all clock edges are aligned, otherwise timing mismatch might arise. While for slow inputs this timing mismatch is not an issue, for higher input frequencies, it may introduce errors. Timing mismatch between the clock edges is inevitable in the clock divider, as each generated clock has a clock edge that is delayed with respect to the clock edge of the previous stage. Hence, this delay should be compensated, which is not trivial to achieve (and probably not very PVT robust). In the design in the next chapter, the clock divider is therefore replaced with a clock gating mechanism, which just skips sample cycles to lower the average sampling rate, rather than changing the sample clock.
- Related to the above issue of changing sampling clocks, the CDAC and the CDAC switches would also benefit from going to a clock gating mechanism for the sampling rate. In the presented design, the sample switch and the CDAC had to be designed such that they were fast enough to operate at the maximum sampling rate and also the leakage that occurs when operating at the lowest clock frequency should be small enough not to introduce distortion. This puts a burden on the CDAC requirements. The clock gating mechanism to be presented in the next chapter can relax these requirements.
- The presented design uses two external reference voltages V_{ref1} and V_{ref2} as threshold voltages that need to span 1 LSB. As these are precision voltages delivered by external source meters (see Fig. 4.17), this is not very practical to use. A differential ADC avoids these extra reference voltages [147]. Hence, the next chapter will implement a differential ADC. This also increases the robustness against common-mode disturbances (such as, e.g., 50-Hz tones).

(continued)

- As pointed out in the section on the CDAC layout, the design uses a "chessboard" DAC layout, but in the end, likely some of the mismatch observed in the DAC was due to "designer" errors, e.g., parasitics of the layout that were not properly matched. An easier layout scheme was adopted in the follow-up chip design to be presented in the next chapter.

Chapter 5
FREYA: An Event-Driven SoC for Spiking End-to-End Classification

This chapter builds further on the level-crossing analog-to-digital converter (LCADC) from the previous chapter. The discrete-time (DT) LCADC introduced in Chap. 4 showed a good power efficiency. In addition, since the LCADC can work in a synchronized fashion, the architecture is easier to integrate in a full sensor system. Furthermore, the synchronous but event-driven nature allows the LCADC to be multiplexed, which is typically not possible with regular asynchronous LCADCs. This allows the LCADC to be used for the readout of multiple multiplexed channels.

To extend the advantage of LCADCs also to the processing part of a system, they can be integrated together with spiking neural networks (SNNs). The event-driven output of LCADCs can be used as the spike train input of SNNs, on which they can perform signal classification. To fully highlight the capabilities of the DT LCADC, this chapter will present the design of an event-driven system-on-chip (SoC), called FREYA. FREYA uses the LCADC as a multichannel analog-to-spike converter (ASC) that provides an input to an on-chip SNN processor that can perform signal classification. As target application, multichannel electrocardiogram (ECG) readout and classification has been chosen.

This chapter is structured as follows. Section 5.1 will discuss some of the challenges of multichannel biomedical systems with on-chip classifiers and will motivate the design presented in this chapter. Sect. 5.2 will highlight the main contributions of the proposed FREYA SoC. In Sect. 5.3, a system overview of the proposed SoC with multichannel ASC and integrated SNN processor will be given. Section 5.4 will describe in detail the multichannel ASC with adaptive-resolution algorithm and the timing of the ASC and will give a detailed circuit implementation of each block. In Sect. 5.5, an end-to-end operation flow for the chip will be described, which details how the ASC settings can efficiently be chosen and how the SNN model should be trained to use the proposed IC in a real-life application. The measurement results of the IC will be presented in Sect. 5.6. First the multichannel and SNN standalone results will be discussed, before giving the

J. Van Assche, G. Gielen, *Event-Driven Circuit Architectures for Scalable and Adaptive Sensor Readout*, Analog Circuits and Signal Processing,
https://doi.org/10.1007/978-3-031-97285-0_5

results of the full system performance for the ECG application. Finally, the section will compare the presented work to the state of the art. Sect. 5.7 will conclude the chapter.

5.1 Biomedical SoC Challenges

This section will briefly summarize the main challenges of biomedical SoCs and will motivate the design choices of the proposed FREYA system. These challenges are schematically depicted in Fig. 5.1.

Conventional biomedical SoCs multiplex multiple readout channels and perform classification in a frame-based manner (see Fig. 5.1a) [8, 69, 136, 194]. This results

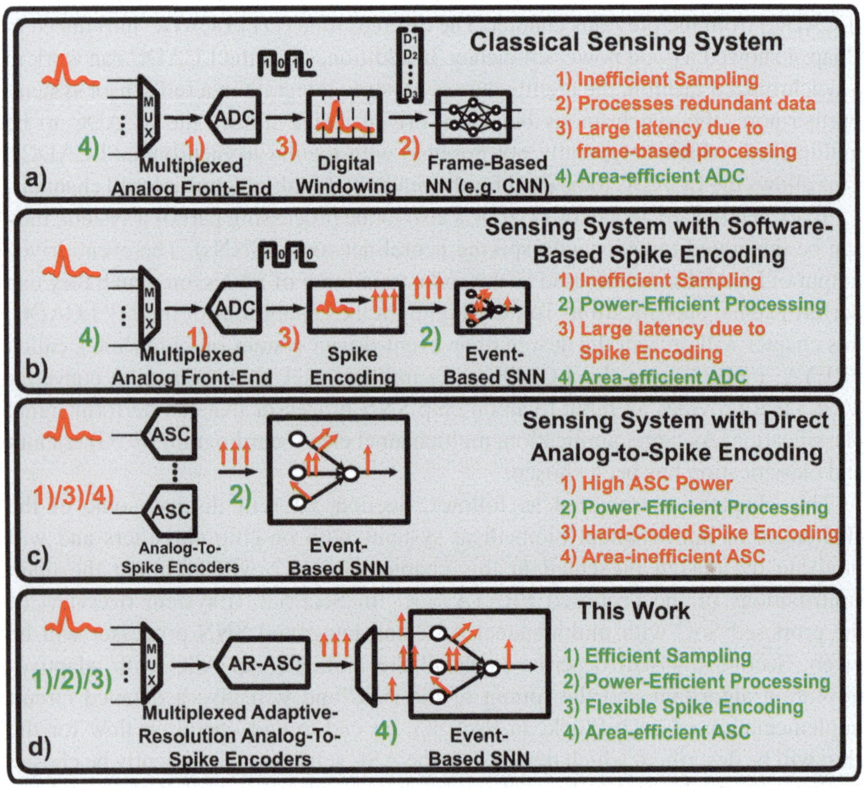

Fig. 5.1 Overview of alternative architectures for biomedical SoCs and their design challenges: (**a**) classical SoC with fixed-rate multichannel ADC, (**b**) event-driven SoC with fixed-rate multichannel ADC and software-based spike encoding, (**c**) event-driven SoC with asynchronous, fixed-threshold ASCs, and (**d**) the proposed event-driven SoC with multichannel, synchronous ASC with adaptive thresholds

in an area-efficient analog-to-digital converter (ADC), since the same hardware can be reused for multiple channels.[1] However, frame-based approaches suffer from large processing latency (>1 s) [69, 194]. Moreover, the digital classifiers used in the frame-based approach, such as, e.g., support vector machines (SVMs) [8], decision trees [136], or convolutional neural networks (CNNs) [69, 151], occupy a large chip area due to the required on-chip memory.[2] Besides the large chip area, large computational network models also consume a significant amount of energy when continuously processing the incoming sensor data. Such systems do not exploit the time-sparse nature of signals like the ECG and likely waste a lot of power in processing redundant data [194].

As alternative to conventional fixed-rate processing, event-based processing SoCs with on-chip SNNs have been proposed (see Fig. 5.1b–d). SNNs can have a smaller processing latency over frame-based neural network topologies due to their spike-based propagation [73, 109]. This event-driven spike propagation also enables an activity-dependent power consumption, which matches well with the time-sparse characteristics of biosignals like ECG [73, 146]. Moreover, area-efficient SNN implementations with dense synapses and neurons have been proposed recently [36, 73], which occupy a smaller on-chip area for the classifier. Event-based SoCs may therefore be an attractive alternative for traditional systems [36, 48, 132, 146]. Several possible architectures for event-based SoCs exist; however, each approach still faces several limitations, as indicated in Fig. 5.1.

SNNs can perform inference on a spike train that is a representation of the input signal [36, 109]. To obtain this spike train, one approach is to first digitize the input signals with a regular fixed-rate ADC, followed by software-based spike encoding. This is shown in Fig. 5.1b [35, 38]. This approach, however, still wastes power by first sampling all sensor data before converting them to spikes. The digital spike conversion also introduces a high latency overhead [38].

Directly converting the analog signals into spikes with a level-crossing sampling (LCS) ASC is a promising alternative (Fig. 5.1c) [48, 94, 132, 133, 167, 168, 171], but the solutions presented till now cannot be multiplexed. The current direct analog-to-spike encoding solutions require a dedicated ASC per recording channel, since conventional ASCs are asynchronous data converters [48, 94, 132]. This results in a large area overhead to implement all the ASCs and in turn limits the channel count that the system can achieve. Moreover, as discussed in the previous two chapters, compared to a regular ADC like a SAR ADC, ASCs can have a power efficiency of up to $10\times$ worse than the regular ADC [48, 132]. Lastly, to increase the accuracy of

[1] Note that this chapter does not focus on the instrumentation amplifier that precedes the ADC, which is typically not multiplexed [136].

[2] As was pointed out in Chap. 2, the size of the CNN memories is relatively large compared to other ML techniques like SVMs and simple classification algorithms. The reduction/optimal allocation of memory size and usage in CNNs is a whole field of research on itself and is out of scope for this work. CNNs can even have a smaller memory footprint compared to the SNNs. Hence, this is why proper benchmarking neural network processors are important and an active field of research [186].

SNNs, it has been shown that encoding the analog signal into spikes with a simple fixed threshold (which is what most LCS ASCs do) results in a lower classification accuracy compared to spike encoding based on an adaptive threshold that can be tuned to the signal of interest [123].

5.2 FREYA: Contributions

To address the issues discussed in the section above, this chapter introduces FREYA, an area- and power-efficient 8-channel event-driven SoC for spiking end-to-end sensing of time-sparse biosignals (see Fig. 5.1d). FREYA offers the following key contributions:

1. An 8-channel time-division-multiplexed LCS ASC that encodes analog input signals into spikes as input for an on-chip SNN, with an area of 0.0067 mm^2/channel. The ASC has a best-case Walden DR FoM of 2.48 pJ/conv.step/channel and a best-case Walden signal-to-noise-and-distortion (SNDR) FoM of 199 fJ/conv.step/channel. The full SoC occupies an area of 0.023 mm^2/channel and has a power consumption of 20.8 μW/channel for the ECG application.
2. The ASC spike encoding algorithm is fully programmable in resolution (4–8 bits) and conversion algorithm (offset and decay parameters) and can flexibly be adjusted to various signals, to achieve the optimal classification accuracy.
3. An on-chip integrated, flexible SNN processor based on a programmable all-to-all crossbar architecture allows for efficient event processing and can easily be reconfigured toward multiple sensing applications. For the ECG QRS-labeling task, a detection accuracy of 98.67% is achieved in software, while the system consumes 13.29 μJ per classification.
4. Based on prior work, in which the relation between the SNN accuracy and the spike encoding scheme of LCS ASCs was explored [123], a custom offline end-to-end training framework for the SoC is introduced. Information-theoretic criteria, such as the corrected Akaike information criterion, can predict for which settings of the LCS ASC the SNN likely reaches the maximum detection accuracy [123]. This technique will be used in the proposed framework to train an SNN model only for a specific subset of the ASC settings. This allows for the fast retraining (69% less training epochs are required compared to a conventional grid-search approach) of the embedded spike encoding algorithm of the ASC and SNN architecture toward new applications or patient-dependent signal variations.

5.3 FREYA: System Overview

The block diagram of the proposed event-driven FREYA SoC is depicted in Fig. 5.2. It consists of two main parts: an adaptive-resolution differential ASC that converts

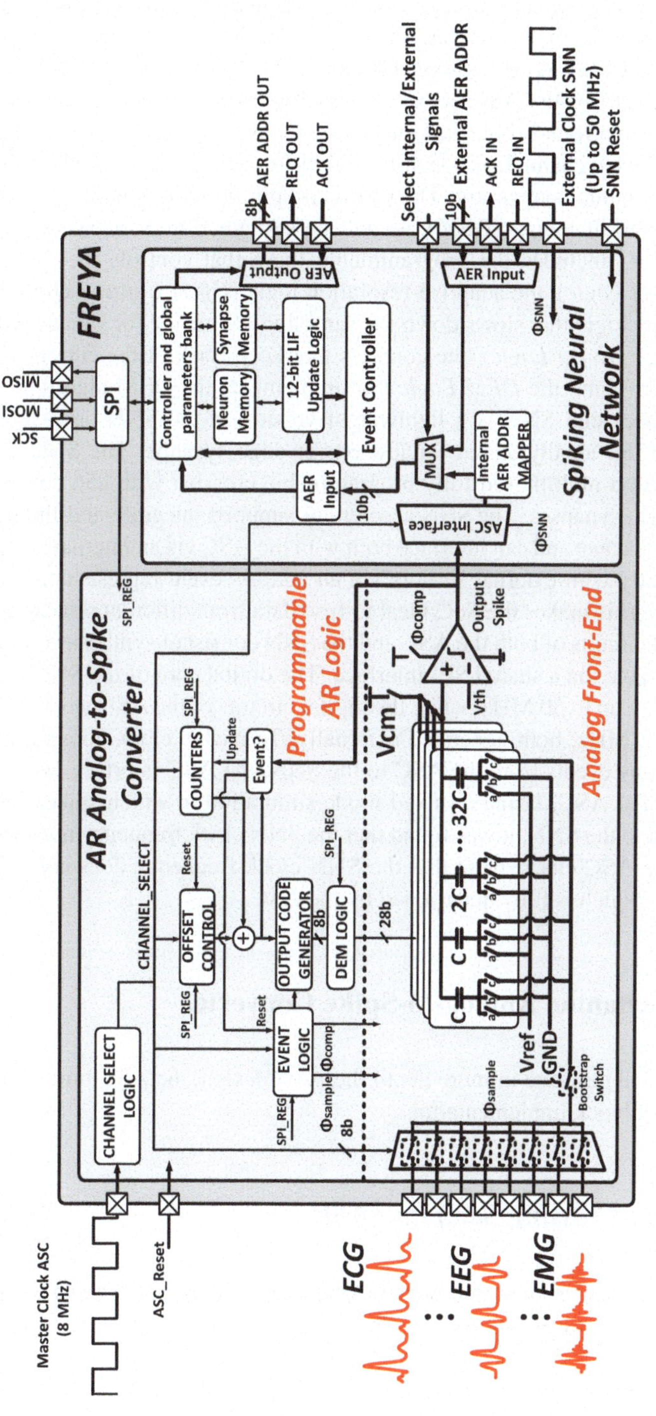

Fig. 5.2 Overview of the proposed event-driven FREYA SoC with adaptive-resolution ASC (drawn here in a single-ended version for reasons of simplicity) and on-chip integrated SNN processor

the incoming analog signal into a spike train and an SNN that performs inference on this spike train. Each channel has a bandwidth of 1000 Hz, targeting biomedical applications, and a full-scale input of 0.7 Vpp.

The analog part of the ASC consists of multiplexer switches, which select the input channels in a round-robin fashion, sampling switches that sample the analog input of the selected channel, a capacitive digital-to-analog converter (CDAC) array, and a single dynamic comparator. The CDAC is split into four sub-arrays to enable dynamic element matching (DEM), as will be explained in Sect. 5.4. The digital part of the ASC includes the programmable logic that controls the multiplexer (*Channel Select Logic*), the adaptive-resolution logic (*Offset Control*, *Output Code Generator*), the logic that slows down the sampling frequency for signals with low input frequency (*Event Logic*), the counters that keep track of the current channel value (*Counters*), and the *DEM Logic* that implements the DEM algorithm of the CDAC. The back-end SNN is a lightweight version of the SNN processor[3] first described in [36], ideally suited for low-power edge systems. The SNN consists of a time-division-multiplexed fully programmable crossbar with 256 neurons and 64k 4-bit weight synapses. The SNN accelerator supports integrate-and-fire neurons with optional leakage and can interface both with the ASC via an internal ASC interface and directly with external sensors via an address-event representation (AER) interface [36]. This makes the SoC ideal to fuse data from different sensor signals.

The digital circuits of both the ASC and the SNN consist of synthesized logic and are programmable via a shared SPI interface. The digital core of the SNN runs at a clock frequency up to 50 MHz, while the digital circuits of the ASC work at a clock frequency of 8 MHz, both generated externally. To enable clock domain crossing and not miss any events from the ASC to the SNN, the ASC interface oversamples the output of the ASC. Using a mixed-mode simulation in which spikes from the ASC were fed to the SNN, it was found that the SNN clock frequency must be $>7\times$ faster than the ASC clock. Note that the SNN clock frequency does not influence the spike rate, which is fully determined by the ASC.

5.4 Multichannel Analog-to-Spike Converter

This section will describe in more detail the ASC design, the ASC timing, and the detailed circuit block implementations.

5.4.1 Level-Crossing Sampling ASC

The previous chapters have introduced event-driven LCS ASCs as an alternative for conventional Nyquist-rate ADCs [48, 94, 132, 133, 167, 168, 171]. Such event-

[3] This processor is available in open source at https://github.com/ChFrenkel/tinyODIN.

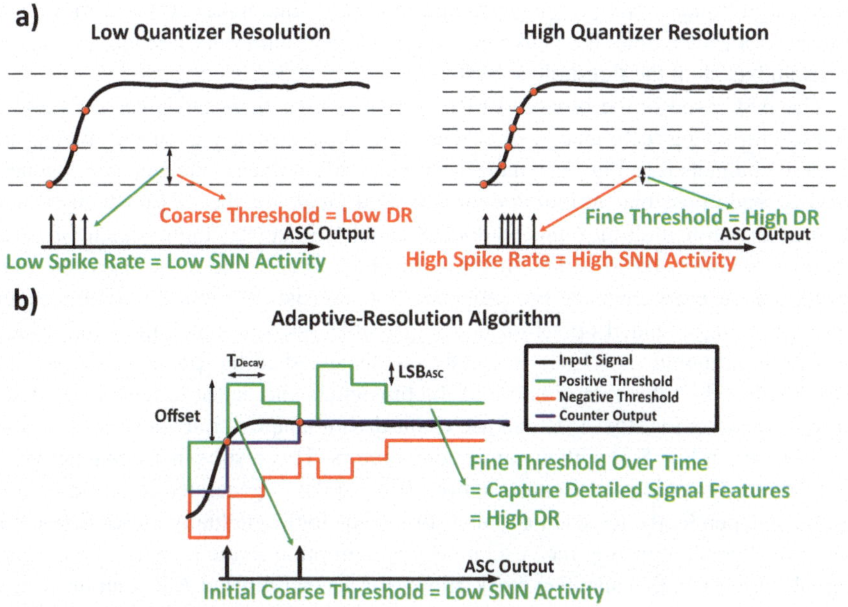

Fig. 5.3 (**a**) Trade-off between spike activity and DR in a conventional LCS ASC and (**b**) the proposed programmable adaptive-resolution level-crossing spike encoding scheme. The programmable parameters of the ASC are the offset, the decay time (T_{Decay}), and the ASC resolution (indicated by the LSB of the ASC, LSB_{ASC})

driven ASCs only take samples whenever the input signal changes beyond a threshold (see Fig. 5.3), which then generates a spike event at the ASC output. These spike events can then be processed directly by SNNs [48, 94, 132, 146].

While interfacing an LCS ASC directly with an SNN allows reducing the latency and the power overhead thanks to the direct analog-to-spike conversion, it still faces circuit challenges compared to traditional fixed-rate sampling systems or software-based spike conversion (see Fig. 5.1a, b). Indeed, since conventional LCS ASCs are asynchronous data converters that are always on, it is difficult to reuse the same hardware in a time-multiplexed fashion, which results in a poor area efficiency for multichannel readout, since a full ASC would be required per readout channel [48, 94, 132]. Moreover, as discussed in Chap. 4, LCS ASCs have a high DC power consumption due to the continuous-time (CT) comparators used to detect a level crossing [48, 94, 132, 171]. Finally, for a given classification task, traditional ASCs face a trade-off between task accuracy and power consumption [146]. If a coarse crossing threshold is used, the ASC generates few spikes, which leads not only to a low SNN dynamic power consumption (Fig. 5.3) but also to a low quantization resolution, and hence a low dynamic range (DR), thereby reducing the SNN classification accuracy [146]. On the other hand, using a finer crossing threshold yields a higher SNN classification accuracy but results in higher spike rates and a higher SNN power consumption [146]. This makes ASCs with a fixed threshold a rather inflexible spike encoding scheme, with only a single degree of freedom based on the threshold selection.

The ASC used in the proposed FREYA SoC solves these challenges. Firstly, a time-multiplexing technique is used where the ASC circuitry is shared among all readout channels (see Fig. 5.2), thereby significantly reducing the area per channel. This is made possible by introducing a system clock for the ASC, which allows to time-division multiplex the readout channels. Since the input signal of each channel is converted to spikes in different clock cycles, the same hardware can be reused for each channel. The proposed SoC consists of eight differential input channels—single-ended operation is also possible—across which the ASC cycles at a programmable rate T_{scan}. Secondly, as discussed in the previous chapter, to fully exploit the time-sparse nature of the biosignals and hence to reduce the ASC power consumption, the ASC can use event-driven adaptive-rate clocking (see the *Event Logic* in Fig. 5.2) with dynamic comparators. This results in a signal activity-dependent, dynamic power consumption. If no spike is generated in a conversion cycle, independently for each channel, the event logic gradually slows down the sampling speed, adjusting the ASC power consumption to the activity of the input signal. Moreover, this adaptive clock allows for a synchronized ASC output, which would be easy to integrate with any subsequent processing/transmitter circuit. Thirdly, each channel of the ASC encodes the input signal by using an adaptive-resolution level-crossing algorithm [133, 171], in which the threshold is initially coarse (i.e., multiple LSBs), before gradually switching to a finer resolution after a programmable amount of cycles T_{Decay} (see Fig. 5.3). If a threshold is crossed, the output of the ASC updates to the value of this threshold and resets the thresholds to the maximum programmable offset value. By employing this adaptive-resolution algorithm for the spike encoding, the usual trade-off between the DR and the ASC spiking activity is improved. By adaptively switching from coarse to fine thresholds, a fine resolution can be achieved that can capture detailed signal features, while at the same time the average spike rate of the ASC, and therefore the SNN dynamic power, is reduced significantly. The adaptive-resolution algorithm and the ASC's programmable resolution implemented on FREYA thus provide three degrees of freedom (the decay time T_{Decay}, the offset, and the ASC resolution), offering much more flexibility for the spike encoding compared to the conventional LCS ASCs.

5.4.2 ASC Timing

The detailed ASC timing is shown in Fig. 5.4. At first, the ASC operates in the *regular mode*, and, at a programmable rate T_{scan}, the ASC cycles over the different readout channels, which is done by the *Channel Select Logic* block. When a channel is selected, the ASC samples the input signal and compares it in four clock cycles (P1–P4) to two thresholds that change adaptively.

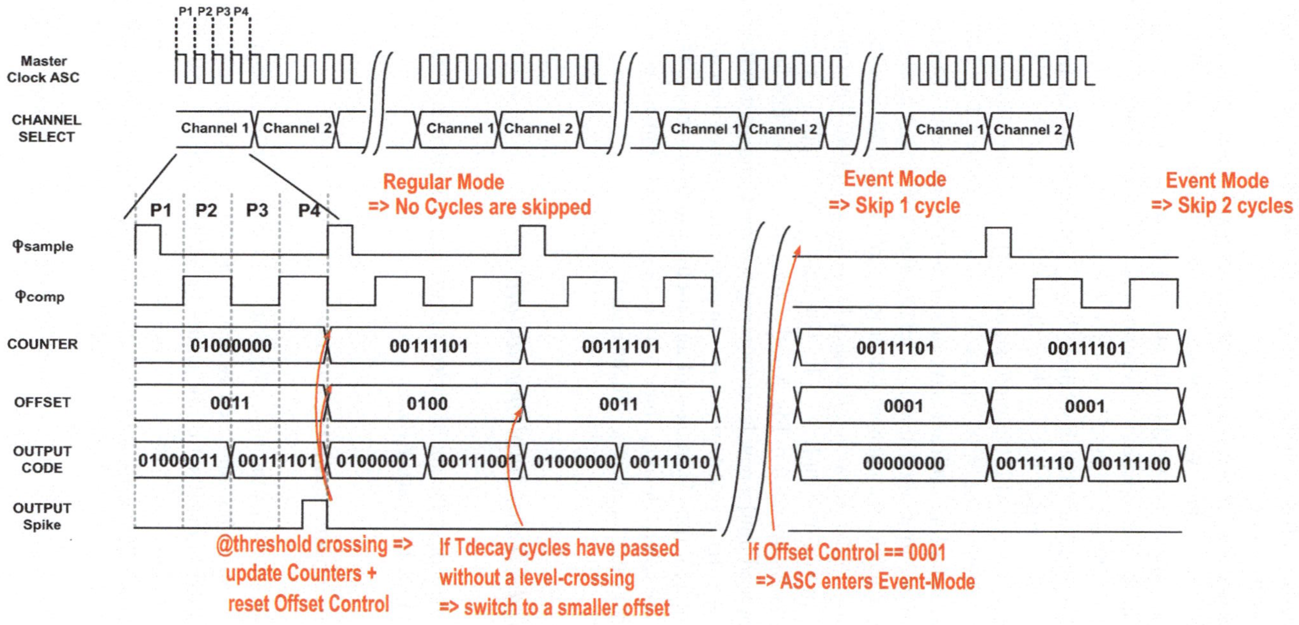

Fig. 5.4 Timing diagram of the proposed adaptive-resolution ASC, implemented on the FREYA SoC. The operation of a single channel is shown in detail; all channels operate independently

The digital code for these threshold levels is generated by the *Offset Control* block, which generates the current offset value, and the *Output Code Generator* block, which takes the offset value and the current value from the *Counters* block and generates a positive threshold during cycles P1 and P2, and a negative threshold during cycles P3 and P4 for the CDACs. If the sampled input signal crosses one of the thresholds, the counter value for the selected channel is updated, and the offset value for that channel is reset to the maximum value, which is programmable via the SPI. Each channel operates independently, and after four cycles, the next channel is selected. Once a channel's offset value is reset, the adaptive-resolution algorithm switches to a finer threshold after T_{Decay} cycles, if no new level crossing occurs. Once the offset value has reached the LSB value for the ASC, which is also programmable via the SPI, the input signal of the readout channel is assumed to be quasi-static. Hence, the ASC enters the *event mode* (see Fig. 5.4), and the ASC starts to skip sampling cycles to save power, until a new level crossing occurs, and the adaptive-resolution algorithm is reset and the ASC enters again the regular mode. In the event mode, the channel select logic keeps cycling over the different channels in the same order. However, if a sample cycle is skipped, the control signals remain low. The amount of sampling cycles that need to be skipped is tracked by the *Event Logic* block. As depicted in Fig. 5.4, the amount of cycles that are skipped is gradually incremented until a new level crossing occurs. Compared to the design in the previous chapter, the event logic skips sample cycles, rather than adjusting the clock frequency in real time by means of an on-chip clock generator. This leads to a simpler and more robust design, since only a single clock domain is used, rather than multiple clocks that need to be synchronized.

5.4.3 Detailed Circuit Implementations

This subsection will describe the detailed circuit implementation of the different building blocks of the proposed ASC.

5.4.3.1 Multiplexer and Sampling Switches

The multiplexer implements the time-division-multiplexing technique of the ASC. It consists of twice eight bootstrapped switches (one for each differential channel), which are connected to the input of the sampling switches of the ASC. Figure 5.5 shows in more detail the multiplexer circuit, the device sizes for the multiplexer, as well as the sampling switches. The bootstrap switches are implemented as described in [116]. Also shown is the logic to control the multiplexer and the sampling switches.

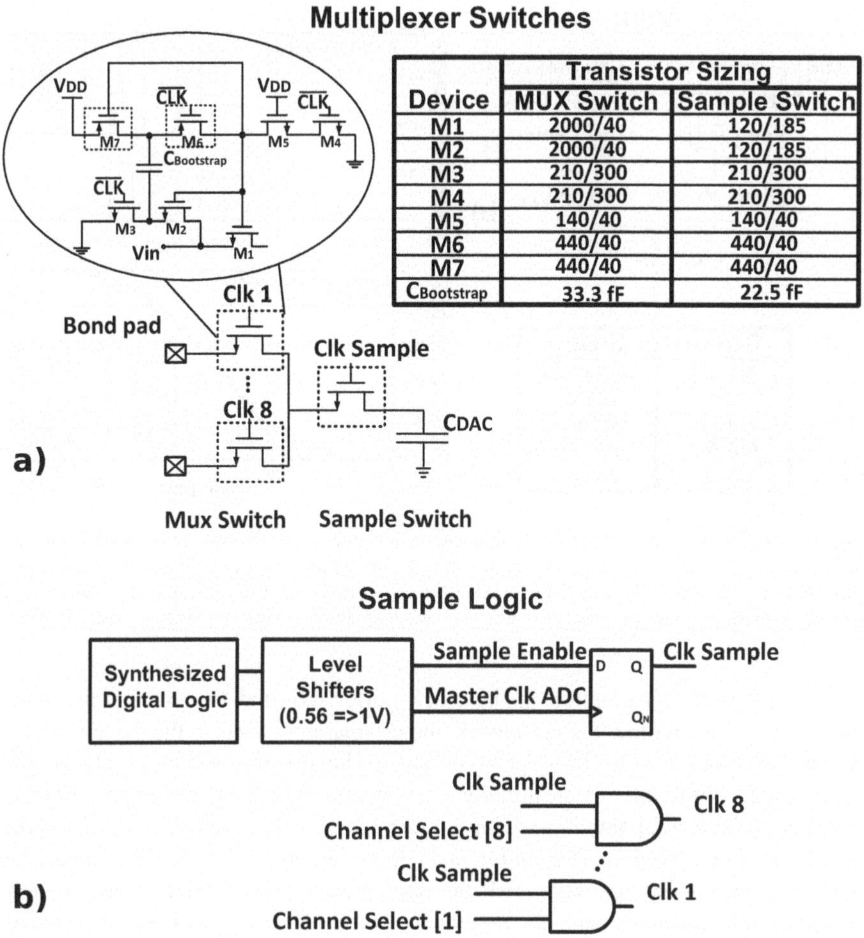

Device	Transistor Sizing	
	MUX Switch	Sample Switch
M1	2000/40	120/185
M2	2000/40	120/185
M3	210/300	210/300
M4	210/300	210/300
M5	140/40	140/40
M6	440/40	440/40
M7	440/40	440/40
$C_{Bootstrap}$	33.3 fF	22.5 fF

Fig. 5.5 (**a**) Single-ended view of the proposed multiplexer circuit and sampling switch, consisting of two rows of bootstrap switches. The device sizes are given in the table. (**b**) Multiplexer and sampling switch control logic

5.4.3.2 Comparator

The previous chapter has proposed to use an adaptive-clocked DT comparator instead of CT comparators like in most ASC implementations. While this can relax the power consumption of the comparators, the required maximum clock frequency for a single channel, given by Eq. (4.1), is relatively large. Hence, increasing this clock frequency to accommodate multiple readout channels would result in a high-frequency ASC clock. Generating such a high-frequency clock may lead to an

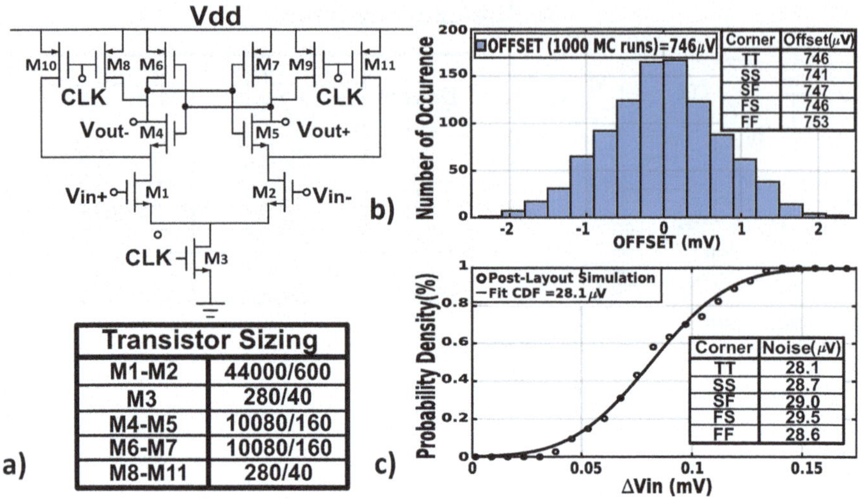

Fig. 5.6 (**a**) StrongARM comparator design and device sizes used. Plotted are (**b**) a 1000-sample Monte Carlo simulation to determine the offset (TT corner) and (**c**) a post-layout transient noise simulation (TT corner). The values in the other process corners are also indicated in the inset

excess in power consumption in the clock buffer required to generate/distribute this clock. To overcome this bottleneck, the comparators used in the design in this chapter are clocked at a lower speed, thanks to the adaptive-resolution algorithm. As this algorithm starts at an initially more coarse threshold, the required clock frequency of the comparators in order not to miss a level crossing can significantly be relaxed, as can be derived from Eq. (4.1). In the proposed design, the comparator is clocked at maximum 4 MHz, half the ASC master clock, which is fast enough to track eight channels with an input bandwidth of 1000 Hz. For comparison, without the adaptive-resolution algorithm, this clock should be 16 $\times$ faster. For the comparator architecture, a StrongARM latch topology [117] has been chosen. The design and its device sizes are given in Fig. 5.6a.

Typically, an LCS ASC uses two comparators to determine if the input signal crosses a threshold above or below the previous level [48, 94, 132, 171]. The difference in input-referred offset of both comparators degrades the ASC linearity (see Sect. 3.2.2). Instead of using power-hungry calibration to compensate the offset [171] or using a common preamplifier to reduce the input-referred offset, which also introduces a power overhead [48], the proposed ASC uses a single comparator for both comparisons, making the comparator offset the same for each comparator decision. In this way, the offset introduces only a DC shift in the quantization levels, which does not degrade the linearity. The comparator offset, however, does reduce the ASC dynamic range. Hence, the comparator devices are sized such that the 3σ offset value is below 0.5 LSB. From post-layout simulations the σ of the offset value

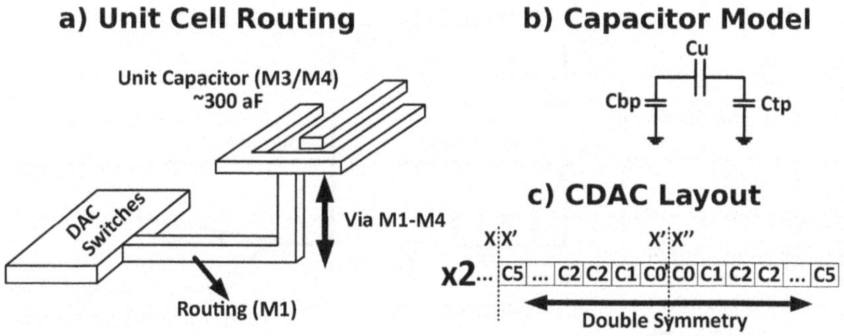

Fig. 5.7 (**a**) Routing of a unit cell of the CDAC, (**b**) the corresponding capacitor model used for circuit simulations, and (**c**) the layout floorplan for a single DAC sub-array

is found to be 746 μV_{rms} (TT corner[4]). The input-referred comparator noise is found to be 28.1 μV_{rms} (TT corner). The simulated values for the offset and noise in the other corners are given in the insets in Fig. 5.6b, c, respectively.

5.4.3.3 CDAC and DEM Logic

For the CDAC design, a custom 300-aF MOMCAP (M3 and M4 metal layers) is used, similar to [46], and depicted in Fig. 5.7. The equivalent capacitor model used for circuit simulations is also given, which was obtained via a parasitic extraction simulation in Siemens Calibre. Cbp and Ctp are 86 and 15 aF, respectively. For the layout of the CDAC, a symmetric layout along the x-axis has been chosen for each single sub-array [46], with double symmetry such that each unit cell consists of two unit capacitors that are placed symmetrically along the axis, averaging out gradients along the x-axis. This layout scheme is kept simple to reduce the fringe capacitance between the routing and the small unit capacitors [46]. For every CDAC, each sub-array is then placed along the y-axis.

[4] While in the context of an academic chip design, designs are usually not optimized to work across different process corners, simulating multiple corners can ensure that a certain design is stable and, hence, provide the designer with some reassurance that the circuit will also work in measurements. Typically, prototype chips are taped out in multi-project wafer (MPW) runs, in which multiple designs are placed together on a single wafer. For these MPW runs, a limited set of wafers is then produced; hence, MPWs are mainly used for research purposes. As only a small set of wafers is produced, there are less concerns about yield, and therefore, wafers coming from an MPW are characterized for their process corners (which is possible by including test structures on the wafer). Hence, the foundry can decide to offer its customers wafers from all different process corners; the default process corner that is offered is the TT corner. If not specifically requested, the foundry throws out wafers that are not in the TT process corner, but obviously this is not possible for industrial-scale chip fabrication in large volumes.

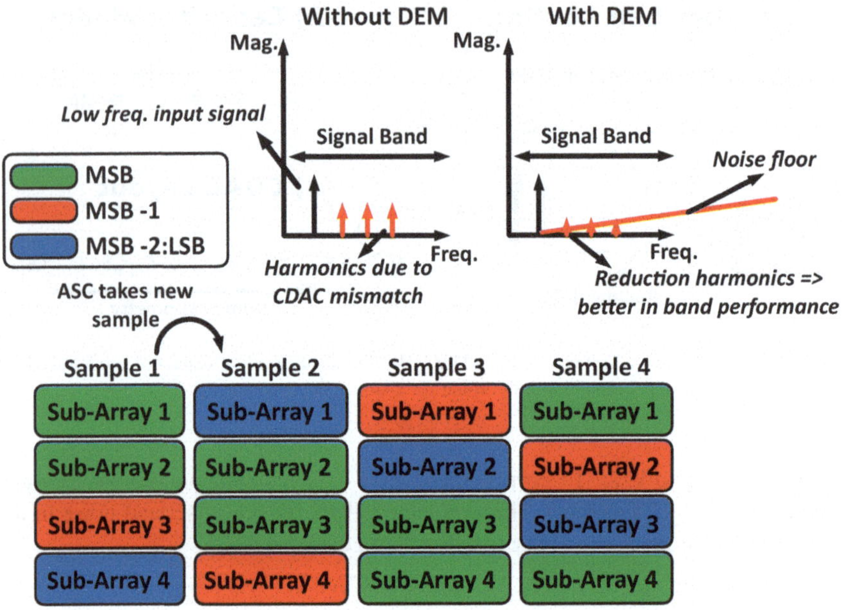

Fig. 5.8 Visual overview of the proposed DEM algorithm for the CDAC sub-arrays and its impact on the ASC distortion

As described in Sect. 3.2.2, LCS ASCs suffer, at low input frequencies, from harmonics that fall into the signal band [177]. These harmonics are caused by both the reconstruction scheme used to interpolate the output signal of the nonuniformly sampling ASC and mismatches in the ASC DAC that generates the threshold values (see also Sect. 3.2.2) [178]. To avoid increasing the sizing of the CDAC to achieve a higher linearity, which would increase the power consumption, a DEM technique is used (Fig 5.8). The 8-bit CDAC is split up into four sub-arrays. In each sample cycle, two sub-arrays are configured as MSB: One sub-array acts as MSB-1 and the other sub-array as a regular binary-weighted DAC. In each new sample cycle, these sub-arrays are shifted in a barrel-shifting fashion, mitigating the harmonics caused by the CDAC mismatch. Another advantage of using DEM is that possible gradient effects, which may not be modeled in a standard PDK, are also averaged out along the y-axis.

5.5 End-to-End Operation of the Event-Driven SoC

FREYA is configurable toward a wide range of time-sparse signal applications. The adaptive-resolution algorithm and the programmable resolution of the ASC allow tailoring the spike encoding to the particular signal of interest. To achieve the highest

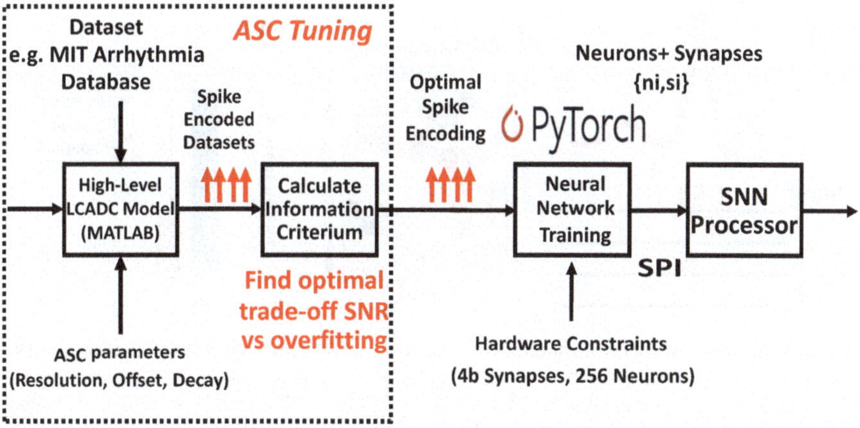

Fig. 5.9 Proposed end-to-end operation flow for the event-driven SoC. The relation between the information criterion and the SNN accuracy has been explored in prior work [123]; the flow presented here applies this insight to train an application end to end

classification accuracy, a custom end-to-end training framework is introduced in this section. It is based on prior work [123], which showed that information-theoretic criteria can predict the settings of an LCS ASC for which the SNN will likely reach the maximum detection accuracy. The framework proposed here uses these insights to deploy FREYA in an end-to-end application, based on the flow depicted in Fig. 5.9. First, a high-level MATLAB model of the ASC generates realistic spike encoding signals for each ASC setting under test; these can be used as input for the SNN training. Doing a brute-force search over all ASC parameters (i.e., the ASC resolution, offset, T_{decay}) for a given application would, however, yield a long optimization time for the SNN model, as a full hyperparameter tuning would be required per ASC setting. For example, for a 10×10 grid search for the offset and T_{decay} parameters, as shown in Fig. 5.10, approximately 10 hours of quantization-aware training would be required per point in the grid on a computer equipped with an NVIDIA V100 GPU. Moreover, simulations show that, depending on the signal (features) of interest, there is a different optimal setting for the ASC, which may even be patient-specific. To avoid a lengthy grid search over all ASC parameters, our training framework calculates, for each spike encoding corresponding to each ASC setting under test, the corrected Akaike information criterion [16] as

$$\mathrm{AIC_c} = N_\mathrm{s}\log\left(\frac{\sum_{n=1}^{N_\mathrm{s}}(\tilde{s}_i - s_i)^2}{N_\mathrm{s}}\right) + 2\kappa + \frac{2\kappa^2 + 2\kappa}{N_\mathrm{s} - \kappa - 1}, \tag{5.1}$$

where N_s is the number of input samples over which the criterion is evaluated (the training dataset), $\tilde{s}_i$ the reconstructed signal from the ASC output, s_i the original input signal, and κ the model complexity, which is the number of nonzero model parameters. Applied to the ASC, it is equal to the spike density of the generated

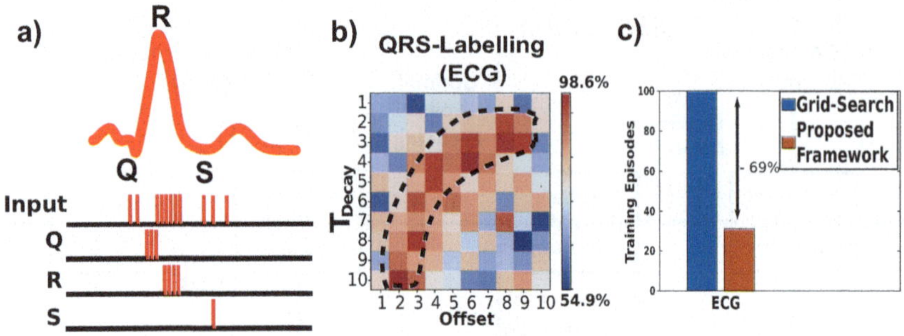

Fig. 5.10 (**a**) Illustrative classification task for the SNN: event-driven QRS-labeling from an ECG input signal. (**b**) The achieved SNN accuracy is plotted as a function of the ASC parameters (the ASC resolution is programmed to be 6 bits). Indicated in the dashed area is the region of optimal ASC performance predicted by the corrected Akaike information criterion. (**c**) This figure indicates the improvement in training time (of 69%) by using the information criterion for training, compared to a full grid search of all ASC parameters

dataset [123]. This corrected Akaike criterion determines a subset of the settings of the ASC for which the ideal trade-off between high DR and overfitting of the model can be reached. Hence, the criterion can be used prior to training the SNN model, to predict if the encoding scheme under test leads to a high or a low SNN accuracy [123]. By calculating this criterion for each ASC setting, the region with minimum information criteria leads to a reduced search space (dashed region in Fig. 5.10b), where the optimal SNN performance is expected to be found. Following this step, for each spike encoding corresponding to an ASC setting in this reduced search space, an SNN model is trained in PyTorch. This training takes into account the hardware constraints of the SNN processor embedded in FREYA, i.e., mainly the limited amount of neurons (256) available and the quantized weights (4 bits signed) [36, 123]. Once the SNN model is trained, the weights and synapses are programmed onto the chip via the SPI interface of FREYA.

As a use case, the QRS-complex labeling task for ECG signals is selected (Fig. 5.10a). Detection of the QRS-complex is often implemented for wearable ECG applications and can be used for, e.g., heart rhythm estimation, arrhythmia detection, etc. [101, 182, 190]. The MIT-BIH Arrhythmia dataset [86] (retrieved via [40]) is used to train the SNN. This dataset contains 48 two-channel ambulatory ECG recordings of 30 minutes each, which were obtained from 47 subjects. Recording 101 is used as the training sequence and recording 201 as the independent test sequence acquired from a different patient.[5] The full 30 minutes of the recording

[5] Recordings 101 and 201 are the first recordings for patient 1 and patient 2, respectively. There was no particular reason for choosing these, other than that they are the first recordings for each patient. It is important to note that each recording contains a large amount of data (30 minutes), which makes the selected recordings more than sufficient for having datasets that capture a wide variety of QRS-complexes. Using more data increases the training time without clear benefits.

are used for training. This leads to many (>1000) data patches for training. It is crucial to note that each recording corresponds to a different patient. If data from both recordings 101 and 102 would be used during training (e.g., by doing a random train/test split), it would make the classification problem easier as the data of both patients are used, and hence, it is not possible to report the test accuracy on new data stemming from a new patient who has not yet been seen during training. Therefore, in line with the best practice rules in machine learning, the data from patient 2 (second recording) are kept as a separate test set, which has never been seen during training, making the test accuracy assessment more fair. The SNN architecture is a recurrent neural network with a single hidden layer of 250 neurons, an input layer of three neurons, and an output layer of three neurons, one for each feature of the QRS-complex, thus three classes. The SNN architecture is trained via back-propagation through time (BPTT) [172] using the SLAYER surrogate gradient technique [140] and the cross-entropy loss function, as conventionally used for classification problems [121]. The Adam optimizer [61] is utilized with learning rate $\eta = 3 \times 10^{-4}$ and decay parameters $\beta_1 = 0.9$ and $\beta_2 = 0.999$, for a total of 100 training epochs with batch size 128. All weights are initialized using the uniform Xavier method [39], following prior SNN training studies such as [121]. During training, the SNN weights are quantized to a 4-bit precision via quantization-aware training [55]. Doing so, a simulated peak classification accuracy of 98.6% is reached for the QRS detection case. By using the proposed training framework, a reduction in NN training epochs of 69% (31 epochs) is obtained compared to the full grid search (100 epochs) for the selected use case (Fig. 5.10c).

Note on Algorithm Versus Circuit Progress

Recent works that combine the sensor readout with on-chip processing (such as the design presented in this chapter) often focus on innovations at multiple levels, i.e., both at the application level and at the circuit level. However, care has to be taken into account when comparing different works. For example, the work in [151] presents an SoC for epilepsy management, which is able to work without the need to be retrained with specific data coming from the patient who will use the SoC.

While this presented SoC makes great progress at the application level and at the circuit level, this SoC (at first glance) appears to be worse compared to other works in the state of the art. Thus, while such works make great progress at the algorithm level, it is difficult to compare such SoC to other works. Another example of such SoC can be found in [131], where the main innovation of this chiplet set is that the decoding of handwritten characters (a complex classification problem with many classes) can fit in a single SoC. This is again a great algorithmic achievement; however, the question remains how to properly compare this type of SoC with other works in terms of circuit performance.

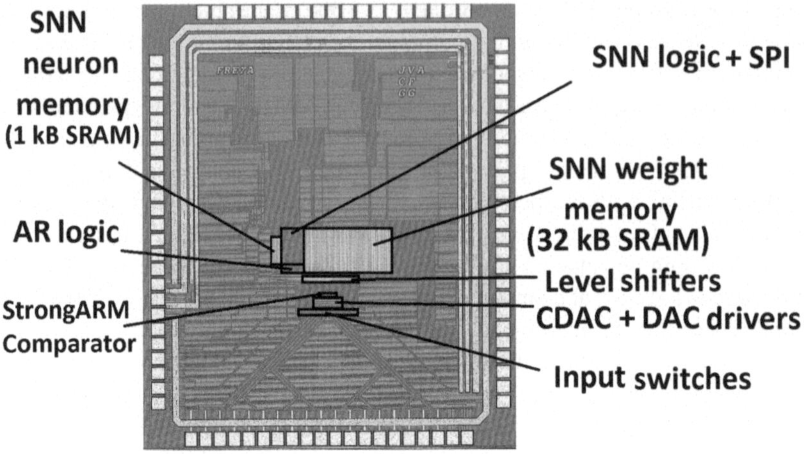

Fig. 5.11 Die photo of the fabricated prototype FREYA IC in 40 nm CMOS

5.6 Measurement Results

The proposed FREYA SoC has been implemented in a 40 nm CMOS technology (see Fig. 5.11) and has a record-low active chip area of 0.023 mm^2/channel (0.184 mm^2 in total, of which the on-chip SRAM takes 70%). The ASC has an area of 0.0067 mm^2/channel.[6] The chip can sample up to eight differential inputs with a maximum bandwidth of 1000 Hz and works at a clock frequency of 8 MHz. The embedded SNN processor can work at a clock frequency up to 50 MHz.

5.6.1 Measurement Setup

A fully automated measurement setup has been developed (see Figs. 5.12 and 5.13). It consists of a Zynq 7020 SoC FPGA, which takes care of the communication with the chip (SPI and AER interfaces), records the data coming from FREYA, and controls the on-PCB components. These consist of a 16-channel DAC (ADI—AD5766) that can provide a multichannel input to the ASC, an optional tunable third-order anti-aliasing filter, on-PCB LDOs (ADI - LT3045) that provide clean reference voltages for FREYA and the board components, instrumentation amplifiers (ADI—AD8237) to amplify small voltages over the current shunt resistors, and an

[6] This is a conservative estimation. The digital circuits for the ASC and SNN are synthesized as one block; hence, the logic circuits from the ASC and SNN are not separable (only the SRAM can be distinguished). For this result, therefore, the area of all digital logic (excluding the SRAM) is taken, which is clearly an overestimation.

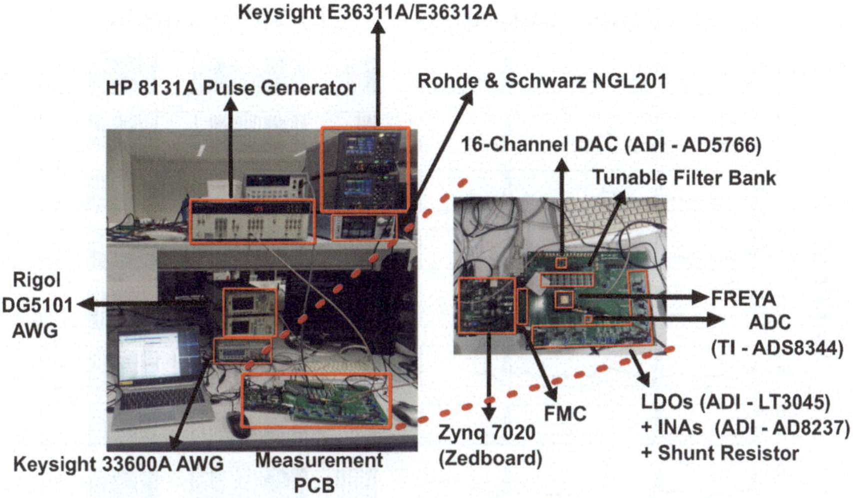

Fig. 5.12 Photo of the measurement setup for FREYA

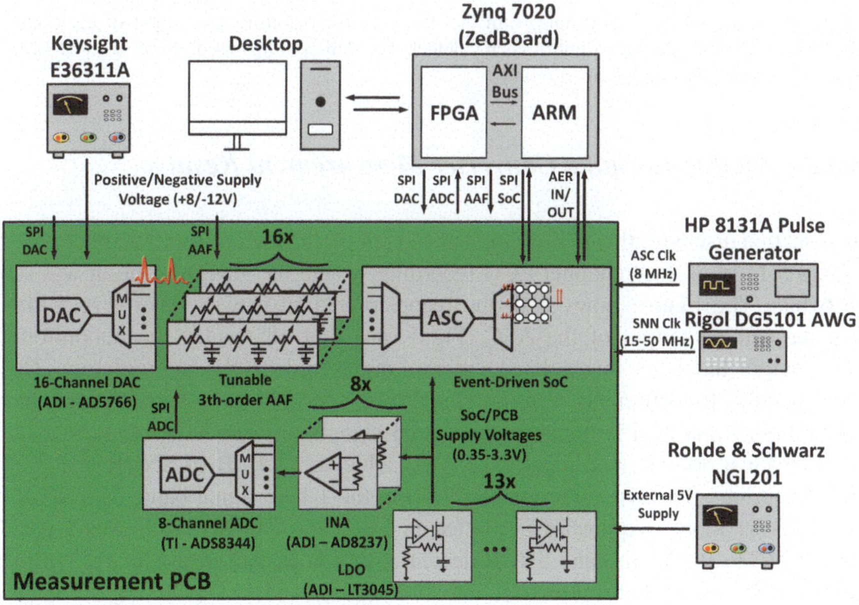

Fig. 5.13 Block diagram of the measurement setup of Fig. 5.12

8-channel ADC (TI—ADS8344) that measures the output of the instrumentation amplifiers to measure the currents of the chip supplies. The clock signals for the ASC and the SNN processor are provided via an external pulse generator (HP 8131A) and an arbitrary waveform generator (Rigol DG5101).

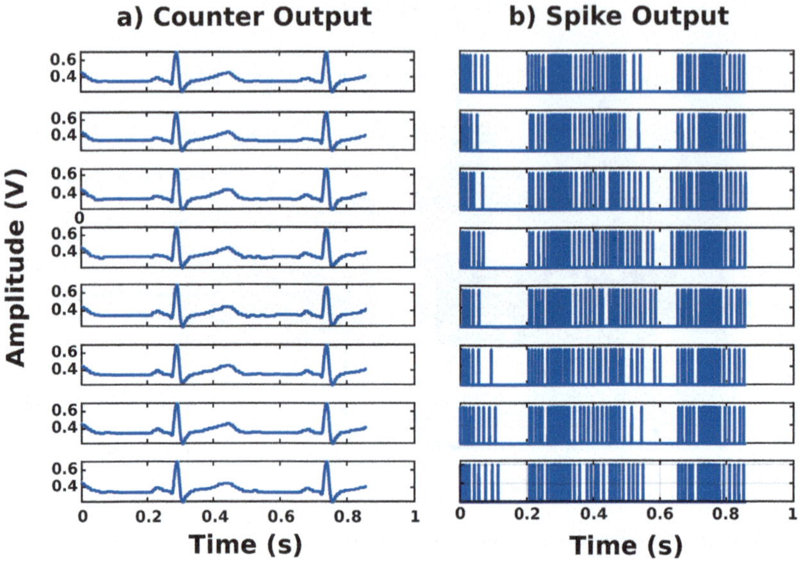

Fig. 5.14 8-channel ECG recording from the ASC: (**a**) the reconstructed output of each ASC channel and (**b**) the spiking output of each channel. The ASC settings used are offset = 1 LSB, $T_{decay} = 0$, and ASC resolution = 6 bits

5.6.2 Analog-to-Spike Converter Measurement Results

This section discusses the measurement results of the standalone multichannel ASC. Figure 5.14 plots an 8-channel ECG recording, where the ASC resolution was set to 6 bits. The figure shows both the reconstructed ECG waveform (Fig. 5.14a) and the spiking output of the ASC (Fig. 5.14b). For all results, the nonuniform ASC output has been interpolated via zero-order-hold (ZOH) interpolation. The DAC mismatch coefficients were corrected by a one-time ramp-based foreground calibration. Figure 5.15 shows the measured signal spectrum of a single readout channel for a 600-Hz sine input. The ASC achieves 71.1 dB of SNDR and 77.9 dB of spurious-free dynamic range (SFDR) in the 1-kHz signal band. The SNDR for different input frequencies is plotted in Fig. 5.16: The ASC achieves a peak SNDR of 74 dB. The in-band SNDR decreases at low frequencies, which is typical for an ASC, when harmonics fall into the signal band [177]. In Fig. 5.17, the measured SFDR of the ASC is shown for different input frequencies, with the DEM enabled and disabled. The DEM technique improves the linearity of the ASC by 7 dB. By applying both calibration and the DEM, the ASC can achieve an even higher linearity of up to 68 dB at low frequencies, which is the theoretical

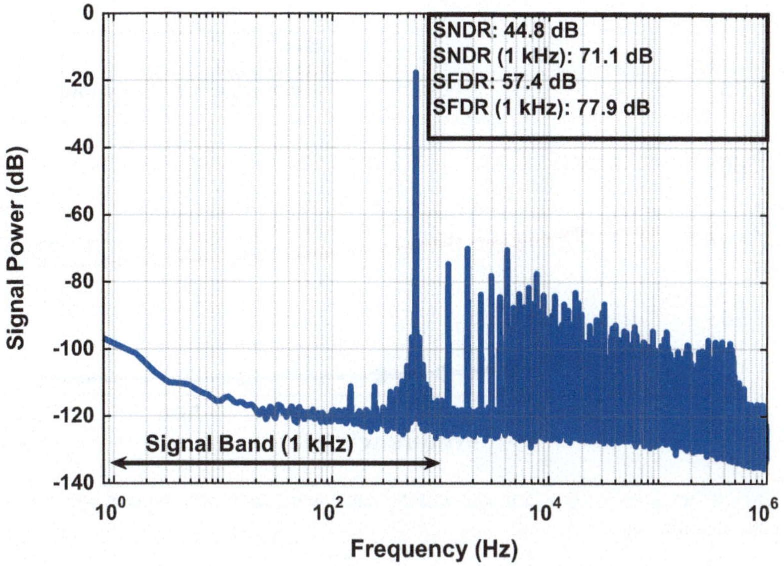

Fig. 5.15 Measured ASC output spectrum of a single channel, for a 700-mVpp 600-Hz sine input

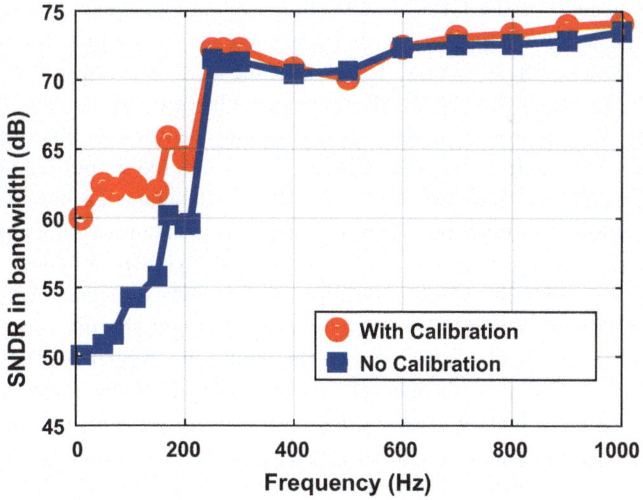

Fig. 5.16 Measured SNDR of the ASC output in the 1-kHz band as a function of the input signal frequency, with and without calibration

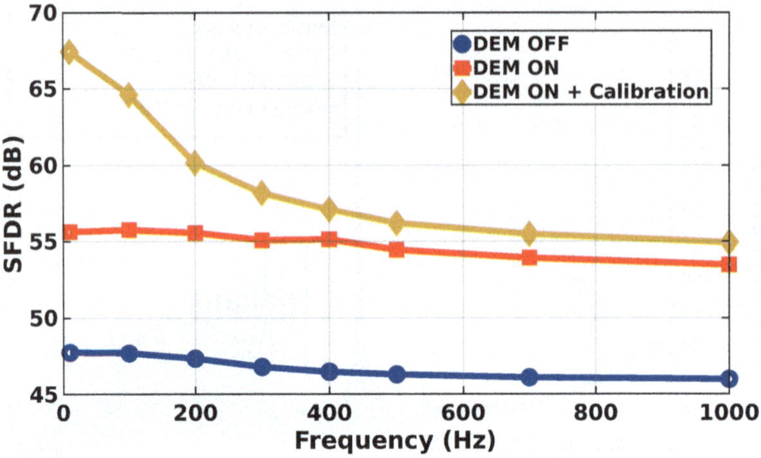

Fig. 5.17 SFDR of the ASC as a function of the input signal frequency, without and with DEM and with calibration on

maximum SFDR for ZOH reconstruction for an 8-bit resolution [177].[7] For higher input frequencies, an SFDR degradation is noted, which has been traced back to a nonoptimally designed bootstrap switch,[8] which introduces distortion at higher input frequencies. For the QRS-classification application, however, this does not matter, as the achieved linearity with DEM (and thus without extra calibration) suffices for the classification task. The INL of the ASC is given in Fig. 5.18. This figure clearly indicates that the DEM improves the linearity of the ASC.

The ASC has a programmable resolution and conversion algorithm, which influence the ADC output data rate. Figure 5.19 shows the output data rate for an ECG signal input for various settings of the ASC; the data rate of a classical Nyquist-rate ADC is plotted as reference. The ASC achieves a data rate that is much lower (e.g., 6$\times$ at 6-bit resolution) than that of a classical Nyquist-rate ADC. Thanks to the dynamic encoding, the power consumption of any processing and/or transmitter blocks that follow the ASC is also relaxed greatly. To highlight this, Fig. 5.20 shows the I/O power of the SoC as a function of the input tone frequency. The I/O power varies from 50 to 190 μW/channel, clearly indicating how the event-driven ASC output can provide a benefit at the system level. Figure 5.20 also plots the power consumption as a function of the input frequency. Note that, with adaptive clocking on, the ASC power consumption scales with the input activity. The ASC consumes between 1.27 and 1.63 μW/channel. The power breakdown of the ASC (for an ECG input and a 1000-Hz sine input) is given in Fig. 5.21. Clearly, the digital logic of

[7] The textbox at the end of the chapter will explain why the DEM algorithm does not fully compensate the mismatch.

[8] This will be discussed in more detail in the textbox at the end of this chapter.

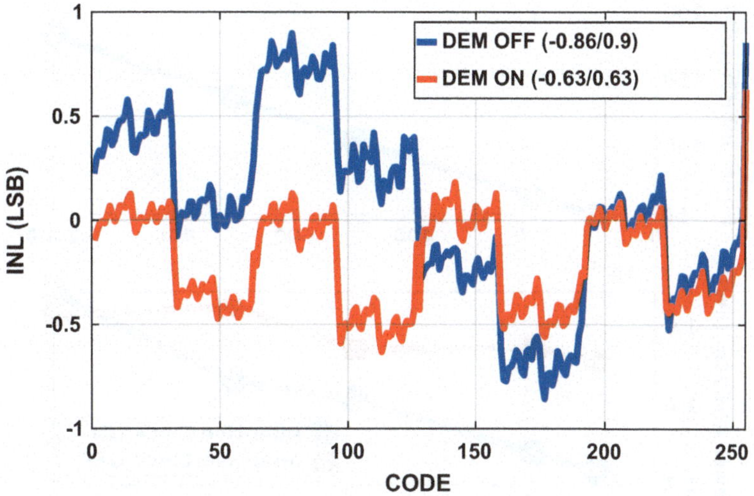

Fig. 5.18 INL of the ASC, with and without the DEM enabled

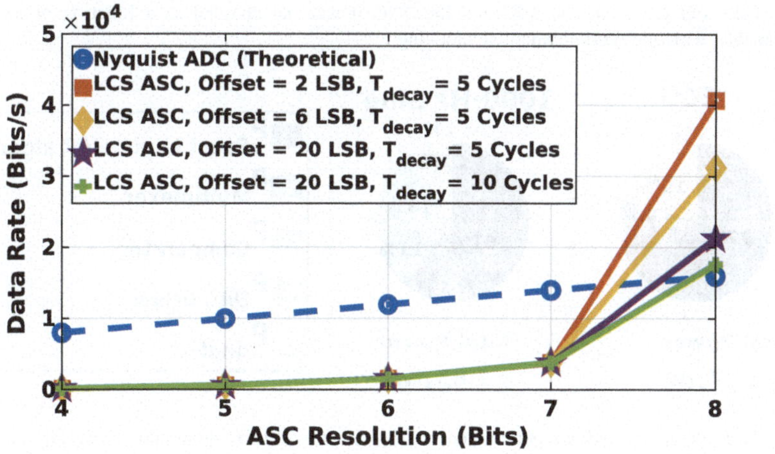

Fig. 5.19 Measured data rate of the ASC for an ECG input signal, plotted as a function of the ASC resolution

the ASC (the event logic + the adaptive-resolution algorithm) consumes the most power. The power of the analog blocks, the comparator ($P_{Comparator}$) and the CDAC (P_{Analog}), scales with the signal activity and only consumes a limited amount of power.

The crosstalk between the recording channels is plotted in Fig. 5.22. For this experiment, all ASC channels have been connected to the ground, except channel 1 that has a full-scale 100-Hz input tone. The signal amplitude in the channel adjacent to the non-grounded channel has been verified, showing a crosstalk of -

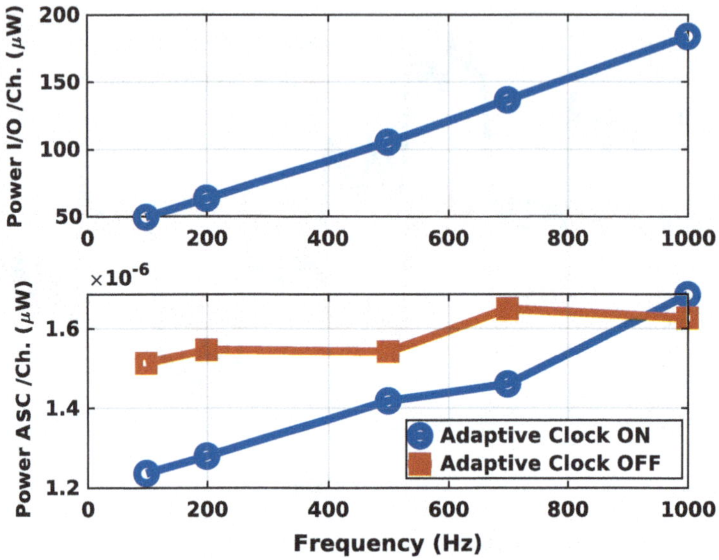

Fig. 5.20 I/O power of the SoC and the ASC power per channel as a function of the input frequency, with and without adaptive clocking

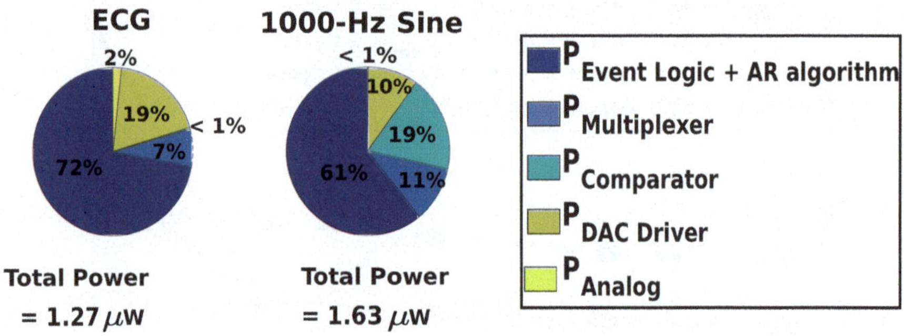

Fig. 5.21 Power breakdown per channel of the ASC, for an ECG input and a 1000-Hz sine input

53.7 dB (Fig. 5.22). To finish the experimental results of the ASC, the presented ASC is compared to the state of the art. In Fig. 5.23, the ASC area/channel is plotted for multiple designs from literature as a function of the DR of the ASCs, clearly showing that the presented ASC significantly reduces the per-channel area, 2× compared to the ASC in Chap. 4.[9] The ASC achieves a minimum Walden SNDR FoM of 199 fJ/conv.step/channel. The DR FoMs introduced in Chap. 4 (see the

[9] Note that since the proposed ASC has differential inputs, it uses two CDACs. Thus, the CDAC array is twice as large compared to a single-ended design. Most designs in literature have a single-ended input.

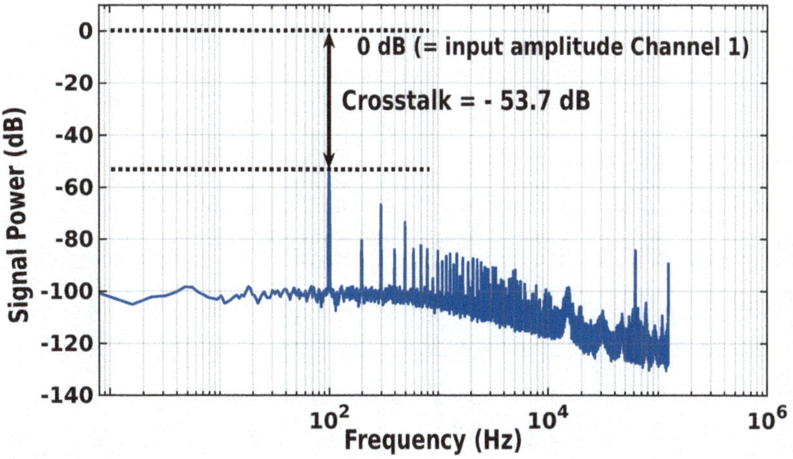

Fig. 5.22 Measured crosstalk in channel 2, when a full-scale, 100-Hz input signal is applied in the adjacent channel (channel 1). The crosstalk is measured by applying an input to channel 1 and grounding all other inputs

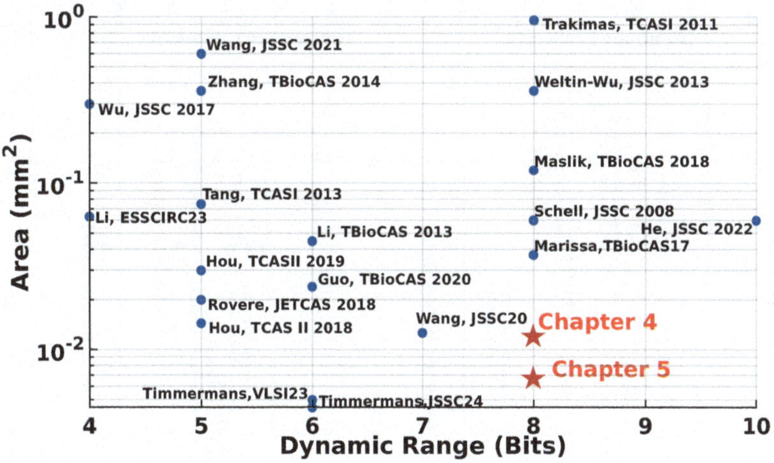

Fig. 5.23 Overview of the area/channel of published ASC designs, as a function of the DR

textbox at the end of the chapter) are also plotted in Fig. 5.24. The ASC achieves a best-/worst-case Walden DR FoM of 2.48/3.18 pJ/conv.step/channel and a best-/worst-case Schreier DR FoM of 138.9/137.8 dB/channel. While the presented ASC's performance may not be the best power efficiency,[10] the goal of this chapter was to highlight how the LCS ASC can be used in a multichannel approach, which

[10] See the textbox at the end of the chapter for a more detailed discussion.

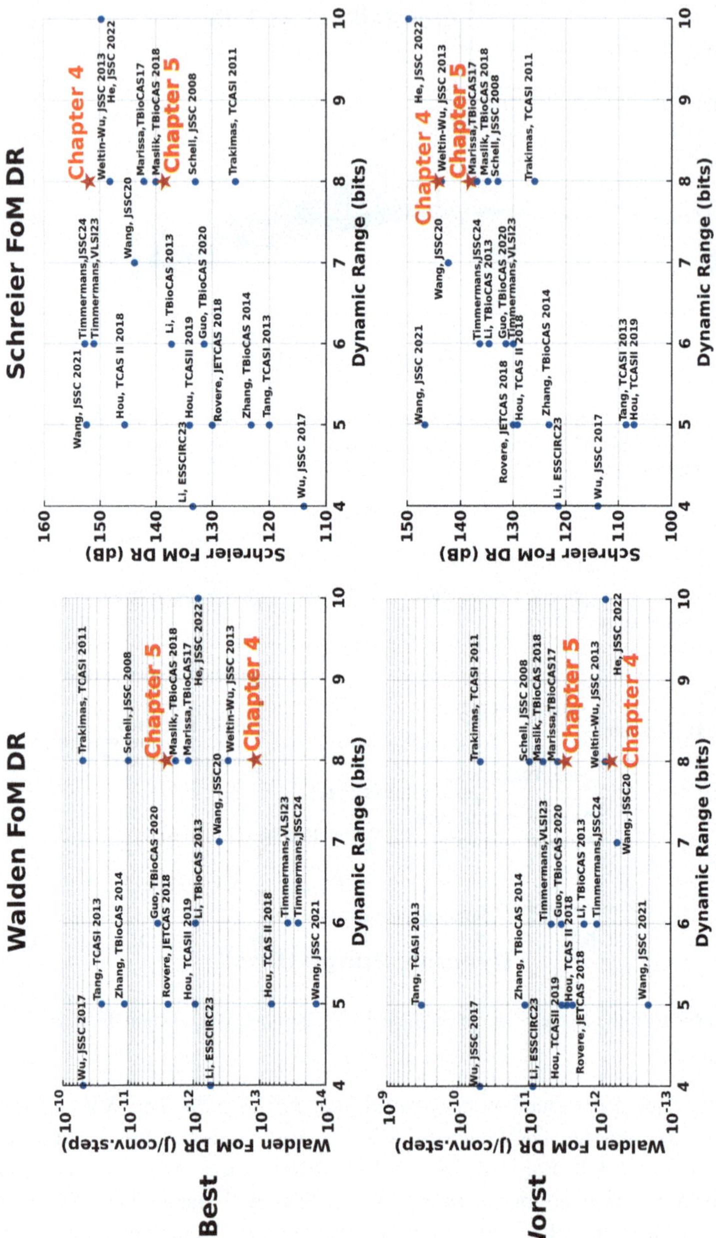

Fig. 5.24 Overview of state-of-the-art ASCs from literature, in the best-case (low input activity) and the worst-case (high input activity) scenario. To calculate the FoMs, the DR is used as the metric for the ASC accuracy

has successfully been demonstrated. In addition, with proper redesign, the power efficiency of the ASC can be made as efficient as the design of the previous chapter.

5.6.3 SNN Measurement Results

The above measurement results have shown how the multichannel ASC provides a spike train for the back-end SNN in a power- and area-efficient manner.

This section will discuss the measurement results for the SNN processor, for the QRS-labeling task presented in Sect. 5.5. Figure 5.25 shows the SNN power consumption (at 0.56 V and 1 V supplies) as a function of the input spike (or event) rate. While a sparser event rate significantly reduces the system power, the leakage and idle power set a lower bound on the benefits of the sparse signal input. Figure 5.26 plots the SNN power consumption as a function of the supply voltage, the amount of input events (shown for 200 events/s and an accelerated input of 10,000 events/s), and the clock frequency, where the programmed network architecture is the one described in Sect. 5.5. For a rate of 200 events/s, which is representative of the ECG use case, the SNN consumes a minimum total power of 155.7 μW. The SNN core can work down to a digital supply of 0.56 V and a clock frequency of 50 MHz without losing functionality. For the SNN clock, the ASC provides a maximum spike rate of 2 MHz. It has been found experimentally that an oversampling ratio between the ASC spike rate and the SNN clock of at least 7× is required. Practically, an SNN clock of 15 MHz has been used.

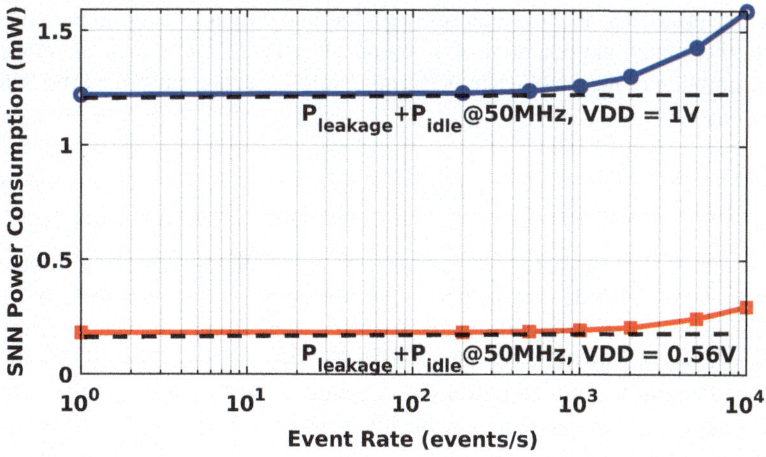

Fig. 5.25 Measured SNN power consumption as a function of the input signal event rate for two different supply voltages

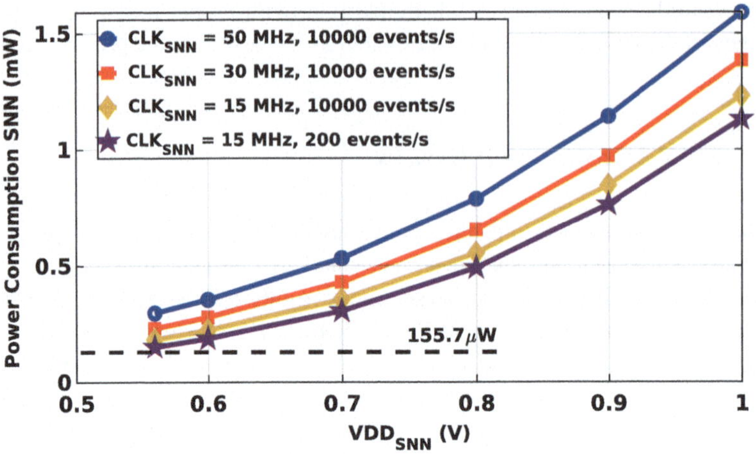

Fig. 5.26 Measured SNN power consumption as a function of the supply voltage, for various input signal event rates and SNN clock frequencies

Design Tip: SNN Energy Efficiency Metrics

To compare the energy efficiency of SNN processors, often metrics like the energy per synaptic operation (E_{SOP}) or the energy per classification (E_{Class}) are used.

E_{SOP} is related to the internal dynamics of the SNN and can be calculated as follows. The synaptic operations (SOPs) are the amount of operations that are performed by all the synapses, e.g., the integration of a weighted spike onto the membrane potential of a neuron [36]. The E_{SOP} is then the power consumption of the SNN when performing inference, normalized by the SOP rate r_{SOP}:

$$E_{\mathrm{SOP}} = \frac{P_{\mathrm{SNN}}}{r_{\mathrm{SOP}}}. \tag{5.2}$$

This metric thus indicates how efficient a single spike propagating through the network is. While this method of calculating the energy efficiency of the SNN seems to be straightforward, there are some things to keep in mind when using this metric to compare different SNN designs. First, depending on the definition, the total power consumption or the dynamic power (which excludes the leakage and the possible idle power consumption) is used for the SNN power consumption [36].

However, often, SNNs consume a non-negligible leakage power, and clocked SNNs also have some added overhead from the clock tree [36]. In

(continued)

the literature [146], the power consumption that was used to calculate the reported metric is not always mentioned. This makes that caution must be taken when comparing different designs. If the leakage power is not reported, it is difficult to assess the reported energy efficiency, as the leakage power can dominate, certainly for real-time applications with low spike activity (e.g., for biosignal applications). Another problem with this metric is that r_{SOP} is not always reported; without mentioning the synaptic activity assumed, the SNN processor's energy efficiency cannot be determined. Moreover, most of the time, the comparison tables note the best-case SOP, which is when the spiking activity is at its maximum (thus, for a high, accelerated input rate). The dynamic power of the SNN is the highest in that case, and the effect of the leakage and the idle power is less outspoken. This leads to a counterintuitive observation: The energy efficiency improves for higher activities. For most biosignal applications, however, the input spike rate is rather low, and therefore also r_{SOP} is lower. Hence, the actual energy efficiency in an application is usually significantly worse compared to the reported best-case efficiency [36].

Another popular metric for the energy efficiency is the amount of energy required to perform a classification for a certain task. For some signals, such as ECG or EEG, certain benchmark tasks (e.g., arrhythmia or seizure onset detection) can be used, such that the SNN accelerators can properly be compared. Moreover, as this classification task is signal specific, the metric can potentially be used to compare the SNN accelerators also to other types of classifiers, like, for example, frame-based CNNs. However, like for the E_{SOP}, the reported E_{Class} can vary greatly, depending on the way how it is measured. E_{Class} is calculated as follows:

$$E_{\text{Class}} = P_{\text{Processor}} \cdot t_{\text{Inference}}, \tag{5.3}$$

with $P_{\text{Processor}}$ the power of the processor during the classification task and $t_{\text{Inference}}$ the time it takes for the classifier to perform a single classification. For sparse signals like the ECG, realistically only a few classifications are done per second. Assuming, for example, a heart rate of 1 beat per second, only three classifications per second have to be done for the QRS-complex detection task. If the real inference time is used, which is determined by the signal, the energy per classification is mainly dominated by, again, the leakage and the idle power. Some works, however, report the energy per classification with an accelerated input; $t_{\text{Inference}}$ is then equal to how fast the SNN makes a decision (i.e., it is equal to the computation latency) [192]. Again, the reported results for this metric and the results in an application can therefore vary wildly, calling for caution to interpret and use the published data.

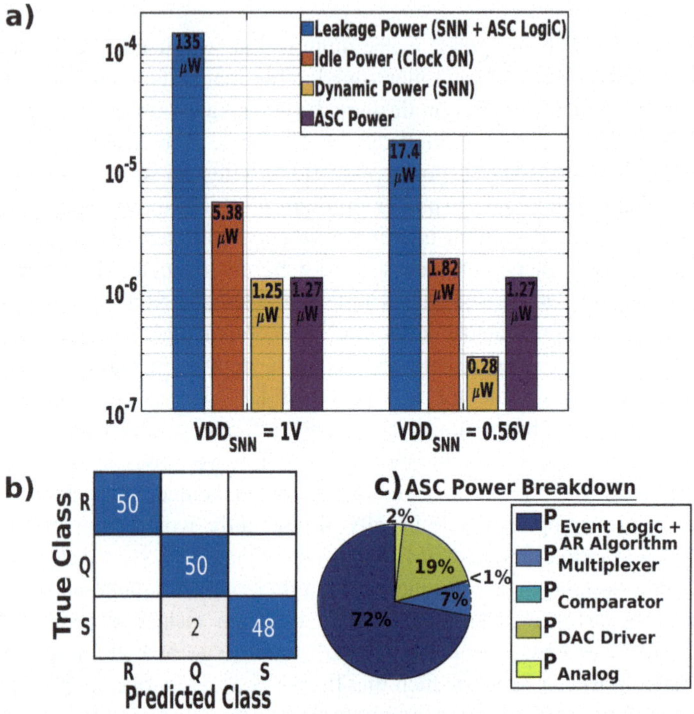

Fig. 5.27 (**a**) The SoC power breakdown per channel for the QRS-labeling application, (**b**) the confusion matrix for the QRS-labeling task, and (**c**) the detailed ASC power breakdown for the QRS-labeling application

5.6.4 System Results

Figure 5.27 shows the system-level results for the QRS-labeling task. To validate the SNN mapping and hardware operation, 50 samples of each class of the classification task discussed in Sect. 5.5 were applied to the FREYA chip via the AER interface. For each sample, the class corresponding to the output neuron that spiked the most was taken as the classification result. These results are summarized in the confusion matrix in Fig. 5.27b. The figure also shows the system power breakdown of the SoC (Fig. 5.27a). For the QRS-classification task at hand, the total system power is 20.8 μW/channel, of which 17.4 μW/channel is static power consumption caused by the leakage in the digital part, which is mostly due to the SRAM macros; 1.82 μW/channel is due to the idle power (clock on), 0.28 μW/channel is due to the dynamic power consumption of the SNN, and 1.27 μW/channel is from the ASC. A more detailed power breakdown for the ASC is provided as well (Fig. 5.27c): At low event rates, the power is mainly consumed by the always-on digital logic of the ASC, as the DAC and the comparator power scale with the signal activity.

For the entire SoC, it is clear that the static power caused by the leakage is the dominant contributor to the power consumption. It is important to note that standard SRAM cells provided by the foundry have been used, which are not optimized for low-power applications. By using custom low-power SRAMs [28, 160], emerging memory technologies such as magnetic RAM (which suffer less from leakage, at the expense of a higher read and write energy cost) or a technology with better leakage performance such as FDSOI, the power efficiency can be improved further.

5.6.5 Comparison with the State of the Art

Table 5.1 lists recent SoCs for biosignal processing besides the designed FREYA SoC. Compared to these other systems, the FREYA SoC achieves at least a 7× reduction in per-channel area, thanks to the proposed time-multiplexed ASC, which is the smallest reported area/channel in the literature (at this time), and the low-footprint digital SNN processor [36]. The SoC thus provides an area-efficient way to process the data from multiple recording channels, which is typically a bottleneck for event-driven SoCs. Compared to the SoCs with frame-based processing [69, 167], the proposed SoC achieves a much smaller latency (80 ms) and area footprint. The SoC consumes 20.8 μW/channel of the total power and 13.29 μJ/classification for the energy per classification.[11] The FREYA SoC has the lowest power consumption/channel for all the systems; only [167] consumes less. However, as [167] only uses a 5-bit LCS ASC, the DR of this LCS ASC is not sufficient, as biomedical systems usually require a DR higher than 7 bits. Note that for always-on systems that need to continuously perform inference (such as the target ECG application), the per-channel power needs to be minimized. The energy/classification of the proposed SoC is a bit on the high side compared to the other works; however, it was taken in more realistic conditions (not at the highest accelerated input), which may explain this higher value (see the textbox above). It can be concluded that the proposed SoC can provide a power-efficient LCS ASC and SNN processor, at a much smaller area footprint.

5.7 Conclusion

This chapter has presented FREYA, an 8-channel event-driven SoC for time-sparse biosignals, consisting of two main parts: a multichannel, time-division-multiplexed ASC and an integrated SNN processor. Thanks to the adaptive system clock and the multiplexing, the proposed ASC can reuse the same hardware for each channel,

[11] To calculate the energy per classification, a $t_{\text{Inference}}$ equal to 80 ms, the SNN latency at 15 MHz, is used, as a spiking input of 200 events/s is chosen.

Table 5.1 Comparison of the FREYA SoC with recently published SoCs for real-time biosignal processing

Reference	TBioCAS23 [69]			JSSC21 [167]		JSSC22 [48]	Nat. Comm 21 [132]	ESSCIRC23 [146]	This work
Technology (nm)	65			180		40	180	40	40
Supply voltage (V)	0.75			0.6		0.9/1/1.1	1.8	0.6/1.1	0.56/1/0.7
Readout channels	1			1		2	8	1	8
Area (mm^2)									
Total	1.74			13.86		0.32	1.42+77.2 (AFE+SNN)***	1.12	0.184
Per channel	1.74			13.86		0.16	0.18+9.65 (AFE+SNN)***	1.12	0.023
Signal	ECG	EEG	EMG	ECG	Audio	ENG	EEG	ECG	ECG
Classification task	Abnormality detection	Seizure detection	Gesture recognition	Abnormality detection	Keyword spotting	De-, Re-, and hyper polarization labeling	High-frequency oscillation detection	Abnormality detection	QRS labeling
Dataset	MIT-BIH. Arr.	Bonn University	NinaPro DB1	MIT-BIH. Arr.	Google Speech	Custom	Clinical iEEG	MIT-BIH. Arr.	MIT-BIH. Arr.

Accuracy (%)	99.3/99.16	99.84	85.13	99.7	99.4	N/A	78	95.31-97.81 (5b-7b ASC)	98.67**
Classes	2/5	2	8	2	2	3	2	5	3
Latency (s)	1.1	1	0.2	0.348		N/A	0.015	N/A	0.08*
Power/Channel (μW)	46.8/86.7	32.1	30	1.68	0.378	25.3	76.75	N/A	20.8
Energy/classification (μJ/class.)	2.25/4.36	2.06	5.25	0.58****	0.13****	N/A	9.21****	0.48-2.25 (5b-7b ASC)	13.29
Including AFE/ADC?	No			Yes		Yes	Yes (Full AFE)	No (Off-Chip ASC used)	Yes
AFE/ADC suited for sensing?	N/A			No (5b)		Yes (10b)	Yes	Yes(7b)	Yes (8b)
Type of processing	Frame-based NN			Event-based AFE + Frame-based NN		Event-based	Event-based	Event-based	Event-based
Multipurpose?	Yes			Yes		No	No	Yes	Yes
Flexible spike encoding	N/A			No		No	No	No	Yes

* Latency measured via applying spikes via the external AER interface, SNN clock running at 15 MHz
Latency $=$ amount of cycles between first input spike and last output spike/ $f_{\mathrm{clk,SNN}}$
** Software accuracy and hardware accuracy verified with 150 samples
*** AFE includes LNA, filters, and ASC
**** Value calculated from data in publication

resulting in an ASC area of only 0.0067 mm^2/channel. The programmable adaptive-resolution algorithm makes it possible to tune the spike encoding to any signal of interest. Signal activity-dependent power consumption is achieved due to the adaptive clocking of the ASC. To efficiently and rapidly tune the ASC and SNN toward a certain application, a custom training framework for the SoC has been presented, which reduces the amount of training epochs required by 69% compared to a full grid search. The framework has been demonstrated for the QRS-labeling task for an ECG input signal, for which a maximum simulated classification accuracy of 98.67% has been reached. Finally, full measurement results of a prototype IC fabricated in a 40 nm CMOS technology have been presented. The chip has a per-channel active area of 0.023 mm^2/channel (0.184 mm^2 in total), which is a 7× improvement over other similar works. For the QRS-labeling task, the system consumes 20.8 μW/channel and 13.29 μJ/classification, with an SNN clock running at 15 MHz (VDD at 0.56V), with 80 ms of latency. Therefore, the presented FREYA SoC paves the way for multichannel, signal-scalable, low-power, and low-area event-driven SoCs for applications that require high-fidelity processing of multiple sensor signals.

Design Errata/Improvements

Like the design of Chap. 5, the design presented in this chapter suffered from some design errors and included some blocks that should be optimized better in a future design:

- Like in the previous chapter, most power consumption of the ASC is consumed by the digital circuits that implement the adaptive-resolution ASC and the event logic. These blocks were designed at the RTL level, but the design mainly focused on functionality, rather than on low power. With some iterations between the RTL implementation and the digital synthesis of the circuits, the power consumption of the digital circuits of the ASC can be reduced further. This would improve the power efficiency of the ASC.
- Due to latch-up DRC violations that only occurred at the MPW top level (e.g., multiple chips combined in 1 GDS file), the bootstrap multiplexer and the sampling switches had to be changed last minute before tape-out. Due to these late changes, there was not enough time to perform a post-layout simulation and to verify that the switches were sufficiently strong to drive the ADC sampling capacitors. This may explain the drop in SFDR of the ASC for higher frequencies. Likely, due to the change in layout, some extra parasitic capacitors have deteriorated the drive strength of the bootstrap capacitor. Large parasitics cause charge sharing between the bootstrap and the parasitic capacitors, which lowers the overdrive voltage of the switch. This theory is somewhat confirmed by the fact that, for a range of 1 Vpp,

(continued)

the ASC suffers from large distortion for the higher input voltages. Thus, the switches are to be redesigned, such that the bootstrap capacitor value can be increased.

- Even though the DEM algorithm increases the SFDR of the ASC, calibration is needed to get to the theoretical maximum SFDR for an 8-bit ASC with ZOH reconstruction. MATLAB simulations showed that the proposed DEM was enough to suppress any random error in the CDAC to this theoretical minimum. This, however, did not reflect in the measurement results. It was noted that the calibration required to obtain the maximum SFDR was not chip specific. The mismatch for each CDAC was very similar from chip to chip. Likely, the remaining error in the CDAC when using DEM may be a systematic error in each sub-DAC, which is not compensated for by the DEM. This could be logical, since, if each sub-DAC has the same systematic error, shuffling each sub-DAC, like the DEM algorithm does, does not remove this error.
- The current SNN model that was used requires that for classification, all the neurons start with a membrane potential equal to 0. After a classification, the neurons should all be reset, but currently this function is not implemented properly. The only possible option is to explicitly set all membrane potentials to zero. But this can only happen via the SPI, which cannot be programmed in real time. This makes that the SNN cannot perform continuous inference on a CT input such as an ECG signal. Thus, either a reset function should be implemented, which is reset after a certain classification window, or an SNN model could be trained that is less sensitive to the membrane potential at the beginning, such that the membrane potentials do not have to be reset after a classification.

Chapter 6
Conclusions

To enable future healthcare applications such as the early detection of heart conditions, closed-loop epilepsy treatment, etc., there is a need to have continuous monitoring (and processing) of vital signs such as the electrocardiogram (ECG) and electroencephalogram (EEG) signals. To enable such applications, low-power sensing devices with wireless connection to the cloud and potentially on-chip processing are required. These edge devices need to operate for long periods of time on a small battery and consume most of their power in data transmission.

This book has focused on the design of integrated sensor readout circuits for biomedical applications. Conventionally, such devices try to reduce the data transmission by means of digital processing of the signals. Digital processing, however, also consumes a significant amount of power. To further improve the efficiency of the sensor readout, this work has explored event-driven readout and processing techniques in the mixed-signal domain.

Chapter 2 has presented a system-level overview of wireless biomedical recording systems, from the electrode to the transmitter. The use case of a 3-lead ECG recording has illustrated that most of the system power is consumed during data transmission. Next, several data reduction techniques have been reviewed, both in the digital and in the mixed-signal domain. While digital techniques also offer the potential for reducing the transmitted data, they often come with substantial power overhead. Conversely, techniques in the mixed-signal domain, such as compressive sensing and event-based processing, have shown promise for minimizing both the processing and the transmission power. Event-based processing, in particular, has been analyzed to offer large potential for creating low-power sensor systems.

Chapter 3 has introduced the event-driven level-crossing ADC (LCADC) architecture, which has variable power consumption and a data rate dependent on the input signal, making it suitable for low-power sensor applications for time-sparse signals like ECG. Two variants of the LCADC, the fixed-window and the floating-window topologies, have been shown as promising circuit implementations for biomedical applications. A MATLAB system-level model has been used to evaluate

J. Van Assche, G. Gielen, *Event-Driven Circuit Architectures for Scalable and Adaptive Sensor Readout*, Analog Circuits and Signal Processing,
https://doi.org/10.1007/978-3-031-97285-0_6

their performance. This model has indicated that, while LCADCs can achieve a high signal-to-noise and distortion ratio (SNDR) with higher-order reconstruction schemes, circuit mismatches might compromise these properties, such that these schemes offer no advantage over simple zero-order-hold reconstruction. Power simulations have been carried out to compare the efficiency of LCADCs and SAR ADCs for various biosignals, showing that LCADCs outperform SAR ADCs for lower resolutions, up to a certain cross-over value (e.g., 6.8 bits for ECG). Most of the power at the ADC level is consumed by the continuous-time (CT) comparators of the LCADC and the time-to-digital converter (TDC). At the system level it has been shown that LCADCs achieve a substantial reduction in transmission power, of two to one order of magnitude for a 5/8-bit resolution, respectively, compared to SAR ADCs, when no retiming is required. Synchronization with a TDC introduces an increase in data rate that offsets the benefits of LCADCs.

Chapter 4 has presented the design of an event-driven LCADC with adaptive clocking. To reduce the large power consumption of the CT comparators, an LCADC topology has been presented in which the CT comparators are replaced by more power-efficient clocked discrete-time (DT) comparators. To keep the advantage of signal-dependent power consumption for time-sparse input signals, an adaptive clocking strategy is used. The adaptive clock is selected by an algorithm (implemented on an off-chip FPGA) that checks the signal activity and, when possible, selects a lower sampling frequency generated from an on-chip clock generator. A prototype LCADC chip has been designed and fabricated in 40 nm CMOS. It achieves an effective number of bits of 10.4 bits, a best-case Walden SNDR/DR figure of merit (FoM) of 138/119.8 fJ/conv.step, and a best-case Schreier SNDR/DR FoM of 158.85/152 dB. The system-level benefits of the topology have also been illustrated: A 30% reduction in data volume for ECG applications has been noted, which results in a large reduction in system power consumption (in particular $3\times$ lower I/O power).

Chapter 5 has presented the FREYA system-on-chip (SoC). This SoC extends the LCADC structure of Chap. 4 to an 8-channel analog-to-spike converter (ASC), which provides inputs to an on-chip spiking neural network (SNN) processor. Thanks to the adaptive system clock, the proposed ASC can time-division multiplex multiple readout channels, while reusing the same hardware for each channel, resulting in an ASC area of only 0.0067 mm^2/channel. The ASC uses a programmable adaptive-resolution algorithm to tune the spike encoding to any signal of interest. To efficiently and rapidly tune the ASC and SNN toward a target application, a custom training framework for the SoC has been presented, which reduces the amount of required training epochs by 69% compared to a full grid search. The framework has been demonstrated for a QRS-labeling task based on an ECG signal, for which a maximum accuracy of 98.67% has been reached. Finally, full measurement results of a second prototype IC, designed and fabricated in a 40 nm CMOS technology, have been presented. The chip has a per-channel active area of 0.023 mm^2/channel, a $7\times$ improvement over other similar works. For the QRS-labeling task, the system consumes 20.8 μW/channel and 13.29 μJ/classification, with an SNN clock running at 15 MHz (VDD at 0.56V), with 80 ms of latency.

The presented prototype ICs in this book improve on the power efficiency, functionality, and area footprint of LCADCs, paving the way for their usage in future low-power biomedical and edge applications.

Appendix A
Zero-Order-Hold Compression and Δ-Encoding

In Chap. 3, Sect. 3.4, the system-level benefits of level-crossing ADCs (LCADCs), e.g., how good do LCADCs reduce the system data rate, have been compared to two light-weight digital compression techniques: zero-order-hold (ZOH) compression and Δ-encoding. This appendix briefly discusses the functioning of both compression methods.

A.1 Zero-Order-Hold Compression

The ZOH compression technique is a lossy compression technique in which, after the ADC has taken a sample, a new sample is only retained if the absolute value of the difference between the current sample and the previous retained sample is larger than Δ*LSB (with Δ an integer value that determines how lossy the compression is). If this difference is smaller than Δ*LSB, the new sample is discarded. This process is continued until a new ADC sample is larger than the programmed Δ*LSB value. To keep track of intermediate discarded values (which are assumed to be equal to the previous sample), a timer is needed that keeps track for how many samples have been skipped. For the analysis in Sect. 3.4, the data rate of this technique is equal to

$$\text{Data Rate ZOH} = F_{\text{dec}} \cdot (N_{\text{ADC}} + N_{\text{Counter}}) \tag{A.1}$$

with F_{dec} the decimated sample rate due to ZOH compression, N_{ADC} the ADC resolution, and N_{Counter} the extra data due to the required timer. The timer resolution is calculated based on the longest time between two new samples and is extracted from a MATLAB simulation for a specific input signal. As ZOH is a lossy compression technique, it introduces an extra compression error on top of the ADC quantization noise. To do a fair comparison, it is assumed that this compression error is equal to the equivalent quantization error of the chosen Δ*LSB. To calculate the

J. Van Assche, G. Gielen, *Event-Driven Circuit Architectures for Scalable and Adaptive Sensor Readout*, Analog Circuits and Signal Processing,
https://doi.org/10.1007/978-3-031-97285-0

compression ratio (CR) for each ZOH set point in Fig. 3.22, the equivalent Nyquist data rate that is used to calculate the CR is assumed for an ADC with an equal quantization error.

A.2 Δ-Encoding

Δ-encoding compresses signals by transmitting the difference between subsequent ADC codes, rather than the ADC codes itself. For signals that do not change much between subsequent samples, for example, the electroencephalogram (EEG), transmitting the difference (the Δ) leads to data compression. The data rate of Δ-encoding can be calculated as

$$\text{Data Rate } \Delta\text{-Encoding} = 2 \cdot f_{\text{Nyquist}} \cdot (N_\Delta + 1), \tag{A.2}$$

where N_Δ is the resolution of the counter that indicates the difference between subsequent ADC codes (the +1 accounts for the sign bit). This counter resolution is extracted from a MATLAB simulation for the input signal. First, the ADC codes for all the samples are calculated, and then the largest difference between two codes is taken. This value is then mapped to a binary value, and N_Δ is taken equal to the rounded-up value of this binary value.

References

1. Analog Devices, AD620 - Low Cost Low Power Instrumentation Amplifier, 2011.
2. iPhone15 - Technical Specifications, 2023.
3. Melexis Company Catalog, 2023. https://media.melexis.com/-/media/files/documents/brochures/company-catalog-melexis.pdf (Accessed: 27-11-2023).
4. ALEA, M. D., SAFA, A., GIACOMOZZI, F., ADAMI, A., TEMEL, I. R., LORENZELLI, L., AND GIELEN, G. A Fingertip-Mimicking 12×16 200m-Resolution e-skin Taxel Readout Chip with per-Taxel Spiking Readout and Embedded Receptive Field Processing. In *2023 IEEE Symposium on VLSI Technology and Circuits (VLSI Technology and Circuits)* (2023), pp. 1–2.
5. ARNRICH, B., MAYORA, O., BARDRAM, J. E., AND TRÖSTER, G. Pervasive Healthcare Paving the Way for a Pervasive, User-Centered and Preventive Healthcare Model. *Methods of Information in Medicine 49*, 1 (2010), 67–73.
6. AZIZ, J. N. Y., ABDELHALIM, K., SHULYZKI, R., GENOV, R., BARDAKJIAN, B. L., DERCHANSKY, M., SERLETIS, D., AND CARLEN, P. L. 256-Channel Neural Recording and Delta Compression Microsystem With 3D Electrodes. *IEEE Journal of Solid-State Circuits 44*, 3 (2009), 995–1005.
7. BERWAL, D., C.R., V., DEWAN, S., C.V., J., AND BAGHINI, M. S. Motion Artifact Removal in Ambulatory ECG Signal for Heart Rate Variability Analysis. *IEEE Sensors Journal 19*, 24 (2019), 12432–12442.
8. BIN ALTAF, M. A., ZHANG, C., AND YOO, J. A 16-Channel Patient-Specific Seizure Onset and Termination Detection SoC with Impedance-Adaptive Transcranial Electrical Stimulator. *IEEE Journal of Solid-State Circuits 50*, 11 (2015), 2728–2740.
9. BINDRA, H. S., ANNEMA, A.-J., LOUWSMA, S. M., AND NAUTA, B. A 0.2–8 MS/s 10 b flexible SAR ADC achieving 0.35–2.5 fJ/conv-step and using self-quenched dynamic bias comparator. In *2019 Symposium on VLSI Circuits* (2019), pp. C74–C75.
10. BINDRA, H. S., LOKIN, C. E., SCHINKEL, D., ANNEMA, A.-J., AND NAUTA, B. A 1.2-V Dynamic Bias Latch-Type Comparator in 65-nm CMOS With 0.4-mV Input Noise. *IEEE Journal of Solid-State Circuits 53*, 7 (2018), 1902–1912.
11. BOHORQUEZ, J. L., YIP, M., CHANDRAKASAN, A. P., AND DAWSON, J. L. A Biomedical Sensor Interface with a sinc Filter and Interference Cancellation. *IEEE Journal of Solid-State Circuits 46*, 4 (2011), 746–756.
12. BOS, T., BADAMI, K., DEHAENE, W., AND VERHELST, M. Architecture optimization for energy-efficient resolution-scalable 8–12-bit SAR ADCs. *Analog Integr Circ Sig Process 97* (2018), 437–448.

J. Van Assche, G. Gielen, *Event-Driven Circuit Architectures for Scalable and Adaptive Sensor Readout*, Analog Circuits and Signal Processing,
https://doi.org/10.1007/978-3-031-97285-0

13. Candes, E., Romberg, J., and Tao, T. Robust uncertainty principles: exact signal reconstruction from highly incomplete frequency information. *IEEE Transactions on Information Theory 52*, 2 (2006), 489–509.
14. Candes, E. J., and Wakin, M. B. An Introduction To Compressive Sampling. *IEEE Signal Processing Magazine 25*, 2 (2008), 21–30.
15. Carlino, M. F., and Gielen, G. Brain Feature Extraction with an Artifact-Tolerant Multiplexed Time-Encoding Neural Frontend for True Real-Time Closed-Loop Neuromodulation. *IEEE Transactions on Biomedical Circuits and Systems* (2023), 1–12.
16. Cavanaugh, J. E. Unifying the derivations for the Akaike and corrected Akaike information criteria. *Statistics & Probability Letters 33*, 2 (1997), 201–208.
17. Chandrakumar, H., and Marković, D. A High Dynamic-Range Neural Recording Chopper Amplifier for Simultaneous Neural Recording and Stimulation. *IEEE Journal of Solid-State Circuits 52*, 3 (2017), 645–656.
18. Chandrakumar, H., and Marković, D. An 80-mVpp Linear-Input Range, 1.6-GΩ Input Impedance, Low-Power Chopper Amplifier for Closed-Loop Neural Recording That Is Tolerant to 650-mVpp Common-Mode Interference. *IEEE Journal of Solid-State Circuits 52*, 11 (2017), 2811–2828.
19. Chang, C.-H., Zahrai, S. A., Wang, K., Xu, L., Farah, I., and Onabajo, M. An Analog Front-End Chip With Self-Calibrated Input Impedance for Monitoring of Biosignals via Dry Electrode-Skin Interfaces. *IEEE Transactions on Circuits and Systems I: Regular Papers 64*, 10 (2017), 2666–2678.
20. Chen, C., Yang, J., Sands, J., Khan, M. A., and Poon, A. S. Next-Generation Wearable Wireless EEG Recorder: The future of accessible neural applications, from mental health monitoring to a noninvasive brain–computer interface. *IEEE Solid-State Circuits Magazine 14*, 4 (2022), 37–50.
21. Chen, F., Chandrakasan, A. P., and Stojanovic, V. M. Design and Analysis of a Hardware-Efficient Compressed Sensing Architecture for Data Compression in Wireless Sensors. *IEEE Journal of Solid-State Circuits 47*, 3 (2012), 744–756.
22. Chen, J., and Ran, X. Deep Learning With Edge Computing: A Review. *Proceedings of the IEEE 107*, 8 (2019), 1655–1674.
23. Chen, M., Chun, H. S., Castro, I. D., Torfs, T., Lin, Q., van Hoof, C., Wang, G., Lian, Y., and van Helleputte, N. A 400 G Input-Impedance Active Electrode for Non-Contact Capacitively Coupled ECG Acquisition With Large Linear-Input-Range and High CM-Interference-Tolerance. *IEEE Transactions on Biomedical Circuits and Systems 13*, 2 (2019), 376–386.
24. Chen, Y., Zhang, X., Lian, Y., Manohar, R., and Tsividis, Y. A Continuous-Time Digital IIR Filter With Signal-Derived Timing and Fully Agile Power Consumption. *IEEE Journal of Solid-State Circuits 53*, 2 (2018), 418–430.
25. Chi, Y. M., Jung, T.-P., and Cauwenberghs, G. Dry-Contact and Noncontact Biopotential Electrodes: Methodological Review. *IEEE Reviews in Biomedical Engineering 3* (2010), 106–119.
26. Chi, Y. M., Maier, C., and Cauwenberghs, G. Ultra-High Input Impedance, Low Noise Integrated Amplifier for Noncontact Biopotential Sensing. *IEEE Journal on Emerging and Selected Topics in Circuits and Systems 1*, 4 (2011), 526–535.
27. Chiang, M., and Zhang, T. Fog and IoT: An Overview of Research Opportunities. *IEEE Internet of Things Journal 3*, 6 (2016), 854–864.
28. Chien, Y.-C., and Wang, J.-S. A 0.2 v 32-kb 10t SRAM with 41 nW standby power for IoT applications. *IEEE Transactions on Circuits and Systems I: Regular Papers 65*, 8 (2018), 2443–2454.
29. Dabbaghian, A., and Kassiri, H. An 8-Channel Ambulatory EEG Recording IC With In-Channel Fully-Analog Real-Time Motion Artifact Extraction and Removal. *IEEE Transactions on Biomedical Circuits and Systems 17*, 5 (2023), 999–1009.
30. De Roose, J., Xin, H., Andraud, M., Harpe, P. J., and Verhelst, M. Flexible and Self-Adaptive Sense-and-Compress for Sub-MicroWatt Always-on Sensory Recording. In

ESSCIRC 2018 - IEEE 44th European Solid State Circuits Conference (ESSCIRC) (2018), pp. 282–285.

31. Dekimpe, R., and Bol, D. ECG Arrhythmia Classification on an Ultra-Low-Power Microcontroller. *IEEE Transactions on Biomedical Circuits and Systems 16*, 3 (2022), 456–466.
32. Dixon, A. M. R., Allstot, E. G., Gangopadhyay, D., and Allstot, D. J. Compressed Sensing System Considerations for ECG and EMG Wireless Biosensors. *IEEE Transactions on Biomedical Circuits and Systems 6*, 2 (2012), 156–166.
33. Donoho, D. Compressed sensing. *IEEE Transactions on Information Theory 52*, 4 (2006), 1289–1306.
34. Fan, Q., Sebastiano, F., Huijsing, J. H., and Makinwa, K. A. A. A 1.8 μ W 60 nV/$\sqrt{}$ Hz Capacitively-Coupled Chopper Instrumentation Amplifier in 65 nm CMOS for Wireless Sensor Nodes. *IEEE Journal of Solid-State Circuits 46*, 7 (2011), 1534–1543.
35. Forno, E., Fra, V., Pignari, R., Macii, E., and Urgese, G. Spike encoding techniques for IoT time-varying signals benchmarked on a neuromorphic classification task. *Frontiers in Neuroscience 16* (12 2022), 1–20.
36. Frenkel, C., Lefebvre, M., Legat, J.-D., and Bol, D. A 0.086-mm^2 12.7-pJ/SOP 64k-Synapse 256-Neuron Online-Learning Digital Spiking Neuromorphic Processor in 28-nm CMOS. *IEEE Transactions on Biomedical Circuits and Systems 13*, 1 (2019), 145–158.
37. Ghaderi, P., Marateb, H. R., and Safari, M.-S. Electrophysiological Profiling of Neocortical Neural Subtypes: A Semi-Supervised Method Applied to in vivo Whole-Cell Patch-Clamp Data. *Frontiers in Neuroscience 12* (2018).
38. Gillet, C., Vincent, A. F., Le Gal, B., and Saïghi, S. Flexible design methodology for spike encoding implementation on FPGA. In *2022 IEEE Biomedical Circuits and Systems Conference (BioCAS)* (2022), pp. 379–383.
39. Glorot, X., and Bengio, Y. Understanding the difficulty of training deep feedforward neural networks. In *International Conference on Artificial Intelligence and Statistics* (2010).
40. Goldberger, A. L., Amaral, L. A. N., Glass, L., Hausdorff, J. M., Ivanov, P. C., Mark, R. G., Mietus, J. E., Moody, G. B., Peng, C.-K., and Stanley, H. E. PhysioBank, PhysioToolkit, and PhysioNet: Components of a New Research Resource for Complex Physiologic Signals. *Circulation 101*, 23 (2000), 215–220. Circulation Electronic Pages: http://circ.ahajournals.org/content/101/23/e215.fullPMID:1085218.
41. Griffith, D. Toward Zero: Power consumption trends in low data rate wireless connectivity. *IEEE Solid-State Circuits Magazine 14*, 4 (2022), 51–60.
42. Guo, W., Kim, Y., Tewfik, A. H., and Sun, N. A Fully Passive Compressive Sensing SAR ADC for Low-Power Wireless Sensors. *IEEE Journal of Solid-State Circuits 52*, 8 (2017), 2154–2167.
43. Hall, D. A., Makinwa, K. A., and Jang, T. Quantifying Biomedical Amplifier Efficiency: The noise efficiency factor. *IEEE Solid-State Circuits Magazine 15*, 2 (2023), 28–33.
44. Harpe, P. A Compact 10-b SAR ADC With Unit-Length Capacitors and a Passive FIR Filter. *IEEE Journal of Solid-State Circuits 54*, 3 (2019), 636–645.
45. Harpe, P. High Precision and Low Power ADCs. In *High Speed/High Performance Data Converters: Metrics, Architectures, and Emerging Topics, ISSCC Short Course* (2022).
46. Harpe, P. J. A., Zhou, C., Bi, Y., van der Meijs, N. P., Wang, X., Philips, K., Dolmans, G., and de Groot, H. A 26 μW 8 bit 10 MS/s Asynchronous SAR ADC for Low Energy Radios. *IEEE Journal of Solid-State Circuits 46*, 7 (2011), 1585–1595.
47. Harrison, R., and Charles, C. A low-power low-noise CMOS amplifier for neural recording applications. *IEEE Journal of Solid-State Circuits 38*, 6 (2003), 958–965.
48. He, Y., Corradi, F., Shi, C., van der Ven, S., Timmermans, M., Stuijt, J., Detterer, P., Harpe, P., Lindeboom, L., Hermeling, E., Langereis, G., Chicca, E., and Liu, Y.-H. An Implantable Neuromorphic Sensing System Featuring Near-Sensor Computation and Send-on-Delta Transmission for Wireless Neural Sensing of Peripheral Nerves. *IEEE Journal of Solid-State Circuits 57*, 10 (2022), 3058–3070.

49. Heikenfeld, J., Jajack, A., Rogers, J., Gutruf, P., Tian, L., T., P., Li, R., Khine, M., Kim, J., Wang, J., and Kim, J. Wearable sensors: modalities, challenges, and prospects. *Lab on a Chip 18* (2018), 217.
50. Hoi-Jun Yoo, C. H. *Bio-Medical CMOS ICs*. (Springer Science+Business Media, LLC, Springer Science+Business Media, LLC, 233 Spring Street, New York, NY 10013, USA), 2011.
51. Hou, Y., Yousef, K., Atef, M., Wang, G., and Lian, Y. A 1-to-1-kHz, 4.2-to-544-nW, Multi-Level Comparator Based Level-Crossing ADC for IoT Applications. *IEEE Transactions on Circuits and Systems II: Express Briefs 65*, 10 (2018), 1390–1394.
52. Hsu, Y.-P., Liu, Z., and Hella, M. M. A 68 dB THD, 0.6 mm^2 Active Area Biosignal Acquisition System With a 40–320 Hz Duty-Cycle Controlled Filter. *IEEE Transactions on Circuits and Systems I: Regular Papers 67*, 1 (2020), 48–59.
53. IDC. Data volume of internet of things (IoT) connections worldwide in 2019 and 2025 , July 27, 2020. https://www.statista.com/statistics/1017863/worldwide-iot-connected-devices-data-size/ (Accessed: 30-11-2023).
54. Ieong, C.-I., Li, M., Law, M.-K., Mak, P.-I., Vai, M. I., and Martins, R. P. A 0.45 V 147–375 nW ECG Compression Processor With Wavelet Shrinkage and Adaptive Temporal Decimation Architectures. *IEEE Transactions on Very Large Scale Integration (VLSI) Systems 25*, 4 (2017), 1307–1319.
55. Jacob, Benoit and Kligys, Skirmantas and Chen, Bo and Zhu, Menglong and Tang, Matthew and Howard, Andrew and Adam, Hartwig and Kalenichenko, Dmitry. Quantization and training of neural networks for efficient integer-arithmetic-only inference. In *Proceedings of the IEEE Conference on Computer Vision and Pattern Recognition (CVPR)* (June 2018).
56. Karam, R., Majerus, S., Bourbeau, D., Damaser, M. S., and Bhunia, S. Ultralow-power data compression for implantable bladder pressure monitor: Algorithm and hardware implementation. In *2016 IEEE Biomedical Circuits and Systems Conference (BioCAS)* (2016), pp. 500–503.
57. Kassiri, H., Salam, M. T., Pazhouhandeh, M. R., Soltani, N., Perez Velazquez, J. L., Carlen, P., and Genov, R. Rail-to-Rail-Input Dual-Radio 64-Channel Closed-Loop Neurostimulator. *IEEE Journal of Solid-State Circuits 52*, 11 (2017), 2793–2810.
58. Kim, D., and Seok, M. A Fully Integrated Digital Low-Dropout Regulator Based on Event-Driven Explicit Time-Coding Architecture. *IEEE Journal of Solid-State Circuits 52*, 11 (2017), 3071–3080.
59. Kim, H., Kim, S., Van Helleputte, N., Artes, A., Konijnenburg, M., Huisken, J., Van Hoof, C., and Yazicioglu, R. F. A Configurable and Low-Power Mixed Signal SoC for Portable ECG Monitoring Applications. *IEEE Transactions on Biomedical Circuits and Systems 8*, 2 (2014), 257–267.
60. Kim, J., Ouh, H., and Johnston, M. L. Multi-Channel Biopotential Acquisition System Using Frequency-Division Multiplexing With Cable Motion Artifact Suppression. *IEEE Transactions on Biomedical Circuits and Systems 15*, 6 (2021), 1419–1429.
61. Kingma, D. P., and Ba, J. Adam: A Method for Stochastic Optimization. *CoRR abs/1412.6980* (2014).
62. Kurchuk, M., and Tsividis, Y. Signal-Dependent Variable-Resolution Clockless A/D Conversion With Application to Continuous-Time Digital Signal Processing. *IEEE Transactions on Circuits and Systems I: Regular Papers 57*, 5 (2010), 982–991.
63. Li, H., Maddox, M., Coin, M. C. W., Buckley, W., Hummerston, D., and Naeem, N. A signal-independent background-calibrating 20b 1MS/S SAR ADC with 0.3 ppm INL. In *2018 IEEE International Solid-State Circuits Conference - (ISSCC)* (2018), pp. 242–244.
64. Li, M., Song, S., Qu, W., Ye, L., Zhao, M., and Tan, Z. A 1.2 V 62.2 dB SNDR SAR-Assisted Event-Driven Clockless Level-Crossing ADC for Time-Sparse Signal Acquisition. In *ESSCIRC 2023- IEEE 49th European Solid State Circuits Conference (ESSCIRC)* (2023), pp. 289–292.

65. Li, Y., Mansano, A. L., Yuan, Y., Zhao, D., and Serdijn, W. A. An ECG Recording Front-End With Continuous-Time Level-Crossing Sampling. *IEEE Transactions on Biomedical Circuits and Systems 8*, 5 (Oct 2014), 626–635.
66. Li, Y., Zhao, D., and Serdijn, W. A. A Sub-Microwatt Asynchronous Level-Crossing ADC for Biomedical Applications. *IEEE Transactions on Biomedical Circuits and Systems 7*, 2 (2013), 149–157.
67. Lichtsteiner, P., Posch, C., and Delbruck, T. A 128× 128 120 dB 15 μs Latency Asynchronous Temporal Contrast Vision Sensor. *IEEE Journal of Solid-State Circuits 43*, 2 (2008), 566–576.
68. Lin, M. P.-H., Hsiao, V. W.-H., Lin, C.-Y., and Chen, N.-C. Parasitic-Aware Common-Centroid Binary-Weighted Capacitor Layout Generation Integrating Placement, Routing, and Unit Capacitor Sizing. *IEEE Transactions on Computer-Aided Design of Integrated Circuits and Systems 36*, 8 (2017), 1274–1286.
69. Liu, J., Fan, J., Zhong, Z., Qiu, H., Xiao, J., Zhou, Y., Zhu, Z., Dai, G., Wang, N., Liu, Q., Xie, Y., Liu, H., Chang, L., and Zhou, J. An Ultra-Low Power Reconfigurable Biomedical AI Processor With Adaptive Learning for Versatile Wearable Intelligent Health Monitoring. *IEEE Transactions on Biomedical Circuits and Systems 17*, 5 (2023), 952–967.
70. Liu, J., Tang, X., Zhao, W., Shen, L., and Sun, N. A 13-bit 0.005-mm2 40-MS/s SAR ADC With kT/C Noise Cancellation. *IEEE Journal of Solid-State Circuits 55*, 12 (2020), 3260–3270.
71. Liu, X., Zhang, M., Xiong, T., Richardson, A. G., Lucas, T. H., Chin, P. S., Etienne-Cummings, R., Tran, T. D., and Van der Spiegel, J. A Fully Integrated Wireless Compressed Sensing Neural Signal Acquisition System for Chronic Recording and Brain Machine Interface. *IEEE Transactions on Biomedical Circuits and Systems 10*, 4 (2016), 874–883.
72. Liu, X., Zhu, H., Zhang, M., Richardson, A. G., Lucas, T. H., and Van der Spiegel, J. Design of a low-noise, high power efficiency neural recording front-end with an integrated real-time compressed sensing unit. In *2015 IEEE International Symposium on Circuits and Systems (ISCAS)* (2015), pp. 2996–2999.
73. Liu, Y., Ma, Y., He, W., Wang, Z., Shen, L., Ru, J., Huang, R., and Ye, L. An 82-nW 0.53-pJ/SOP Clock-Free Spiking Neural Network With 40-μs Latency for AIoT Wake-Up Functions Using a Multilevel-Event-Driven Bionic Architecture and Computing-in-Memory Technique. *IEEE Transactions on Circuits and Systems I: Regular Papers 70*, 8 (2023), 3075–3088.
74. Lohit, S., Kulkarni, K., and Turaga, P. Direct inference on compressive measurements using convolutional neural networks. In *2016 IEEE International Conference on Image Processing (ICIP)* (2016), pp. 1913–1917.
75. Lustig, M., Donoho, D. L., Santos, J. M., and Pauly, J. M. Compressed Sensing MRI. *IEEE Signal Processing Magazine 25*, 2 (2008), 72–82.
76. Mamaghanian, H., Khaled, N., Atienza, D., and Vandergheynst, P. Compressed Sensing for Real-Time Energy-Efficient ECG Compression on Wireless Body Sensor Nodes. *IEEE Transactions on Biomedical Engineering 58*, 9 (2011), 2456–2466.
77. Mamaghanian, H., Khaled, N., Atienza, D., and Vandergheynst, P. Compressed Sensing for Real-Time Energy-Efficient ECG Compression on Wireless Body Sensor Nodes. *IEEE Transactions on Biomedical Engineering 58*, 9 (2011), 2456–2466.
78. Manyika, J., Chui, M., Bisson, P., Woetzel, J., Dobbs, R., Bughin, J., and Aharon, D. The internet of things: mapping the value beyond the hype.
79. Marisa, T., Niederhauser, T., Haeberlin, A., Wildhaber, R. A., Vogel, R., Goette, J., and Jacomet, M. Pseudo Asynchronous Level Crossing ADC for ECG Signal Acquisition. *IEEE Transactions on Biomedical Circuits and Systems 11*, 2 (2017), 267–278.
80. Mark, J., and Todd, T. D. A Nonuniform Sampling Approach to Data Compression. *IEEE Transactions on Communications 29*, 1 (1981), 24–32.
81. Martínez-Nuevo, P., Patil, S., and Tsividis, Y. Derivative Level-Crossing Sampling. *IEEE Transactions on Circuits and Systems II: Express Briefs 62*, 1 (2015), 11–15.

82. MASLIK, M., LIU, Y., LANDE, T. S., AND CONSTANDINOU, T. G. Continuous-Time Acquisition of Biosignals Using a Charge-Based ADC Topology. *IEEE Transactions on Biomedical Circuits and Systems 12*, 3 (2018), 471–482.
83. MEIJER, G. C. *Smart Sensor Systems*. John Wiley & Sons, Ltd, 2008.
84. MOIS, G., FOLEA, S., AND SANISLAV, T. Analysis of Three IoT-Based Wireless Sensors for Environmental Monitoring. *IEEE Transactions on Instrumentation and Measurement 66*, 8 (2017), 2056–2064.
85. MONTEVECCHI, F., STICKLER, T., HINTEMANN, R., AND HINTERHOLZER, S. Energy-efficient cloud computing technologies and policies for an eco-friendly cloud market. *European Commission, final study report, Vienna, report commissioned by the Directorate-General for Communications Networks, Content and Technology* (2020).
86. MOODY, G., AND MARK, R. The impact of the MIT-BIH Arrhythmia Database. *IEEE Engineering in Medicine and Biology Magazine 20*, 3 (2001), 45–50.
87. MOORE, G. E. Cramming More Components onto Integrated Circuits. *Electronics* (April 19, 1965), 114–117.
88. MORA LOPEZ, C., PUTZEYS, J., RADUCANU, B. C., BALLINI, M., WANG, S., ANDREI, A., ROCHUS, V., VANDEBRIEL, R., SEVERI, S., VAN HOOF, C., MUSA, S., VAN HELLEPUTTE, N., YAZICIOGLU, R. F., AND MITRA, S. A Neural Probe With Up to 966 Electrodes and Up to 384 Configurable Channels in 0.13 μm SOI CMOS. *IEEE Transactions on Biomedical Circuits and Systems 11*, 3 (2017), 510–522.
89. MOURRANE, S., LARRAS, B., CLERC, S., CATHELIN, A., AND FRAPPÉ, A. A 291 nW Real-Time Event-Driven Spectrogram Extraction unit in 28 nm FD-SOI CMOS for Keyword Spotting Application. In *ESSCIRC 2023- IEEE 49th European Solid State Circuits Conference (ESSCIRC)* (2023), pp. 341–344.
90. MPITZIOPOULOS, A., GAVALAS, D., KONSTANTOPOULOS, C., AND PANTZIOU, G. A survey on jamming attacks and countermeasures in WSNs. *IEEE Communications Surveys & Tutorials 11*, 4 (2009), 42–56.
91. MULLER, R., GAMBINI, S., AND RABAEY, J. M. A 0.013 mm^2, 5 μW, DC-Coupled Neural Signal Acquisition IC With 0.5 V Supply. *IEEE Journal of Solid-State Circuits 47*, 1 (2012), 232–243.
92. MURMANN, B. ADC Performance Survey 1997–2023. [Online]. Available: https://github.com/bmurmann/ADC-survey.
93. MURMANN, B. The Race for the Extra Decibel: A Brief Review of Current ADC Performance Trajectories. *IEEE Solid-State Circuits Magazine 7*, 3 (2015), 58–66.
94. NARAYANAN, S., CARTIGLIA, M., RUBINO, A., LEGO, C., FRENKEL, C., AND INDIVERI, G. SPAIC: A sub-W/Channel, 16-Channel General-Purpose Event-Based Analog Front-End with Dual-Mode Encoders. In *2023 IEEE Biomedical Circuits and Systems Conference (BioCAS)* (2023), pp. 1–5.
95. NG, K. A., AND XU, Y. P. A Compact, Low Input Capacitance Neural Recording Amplifier. *IEEE Transactions on Biomedical Circuits and Systems 7*, 5 (2013), 610–620.
96. OECD. *Health at a Glance 2023*. 2023.
97. OMNIVISION TECHNOLOGIES. OS04D10 - 4 megapixel product brief, 2023. https://www.ovt.com/wp-content/uploads/2023/03/OS04D10-PB-v1.0-WEB.pdf (Accessed: 30-11-2023).
98. OTTATI, F., GAO, C., CHEN, Q., BRIGNONE, G., CASU, M. R., ESHRAGHIAN, J. K., AND LAVAGNO, L. To Spike or Not to Spike: A Digital Hardware Perspective on Deep Learning Acceleration. *IEEE Journal on Emerging and Selected Topics in Circuits and Systems 13*, 4 (2023), 1015–1025.
99. PAMULA, V. R., VALERO-SARMIENTO, J. M., YAN, L., BOZKURT, A., HOOF, C. V., HELLEPUTTE, N. V., YAZICIOGLU, R. F., AND VERHELST, M. A 172 μW Compressively Sampled Photoplethysmographic (PPG) Readout ASIC With Heart Rate Estimation Directly From Compressively Sampled Data. *IEEE Transactions on Biomedical Circuits and Systems 11*, 3 (2017), 487–496.

100. Pamula, V. R., Verhelst, M., Van Hoof, C., and Yazicioglu, R. F. A 17 nA, 47.2 dB dynamic range, adaptive sampling controller for online data rate reduction in low power ECG systems. In *2016 IEEE Biomedical Circuits and Systems Conference (BioCAS)* (2016), pp. 272–275.
101. Pan, J., and Tompkins, W. J. A Real-Time QRS Detection Algorithm. *IEEE Transactions on Biomedical Engineering BME-32*, 3 (1985), 230–236.
102. Pareschi, F., Albertini, P., Frattini, G., Mangia, M., Rovatti, R., and Setti, G. Hardware-Algorithms Co-Design and Implementation of an Analog-to-Information Converter for Biosignals Based on Compressed Sensing. *IEEE Transactions on Biomedical Circuits and Systems 10*, 1 (2016), 149–162.
103. Patil, S., Ratiu, A., Morche, D., and Tsividis, Y. A 3–10 fJ/conv-step Error-Shaping Alias-Free Continuous-Time ADC. *IEEE Journal of Solid-State Circuits 51*, 4 (2016), 908–918.
104. Pavan, S., Manivannan, S., and Basavaraj, N. Analysis and Design of Wideband Filtering ADCs Using Continuous-Time Pipelining. *IEEE Journal of Solid-State Circuits 59*, 1 (2024), 268–281.
105. Pelgrom, M. *Analog-to-Digital Conversion*. Springer Cham, Springer International Publishing Switzerland 2022, 2022.
106. Pelgrom, M., Duinmaijer, A., and Welbers, A. Matching properties of MOS transistors. *IEEE Journal of Solid-State Circuits 24*, 5 (1989), 1433–1439.
107. Peng, S.-Y., Lee, Y.-H., Wang, T.-Y., Huang, H.-C., Lai, M.-R., Lee, C.-H., and Liu, L.-H. A Power-Efficient Reconfigurable OTA-C Filter for Low-Frequency Biomedical Applications. *IEEE Transactions on Circuits and Systems I: Regular Papers 65*, 2 (2018), 543–555.
108. Pholpoke, B., Songthawornpong, T., and Wattanapanitch, W. A Micropower Motion Artifact Estimator for Input Dynamic Range Reduction in Wearable ECG Acquisition Systems. *IEEE Transactions on Biomedical Circuits and Systems 13*, 5 (2019), 1021–1035.
109. Plagwitz, P., Hannig, F., Teich, J., and Keszocze, O. To Spike or Not to Spike? A Quantitative Comparison of SNN and CNN FPGA Implementations, 2023.
110. Pochet, C., Huang, J., Mercier, P., and Hall, D. A. A 174.7-dB FoM, 2nd-Order VCO-Based ExG-to-Digital Front-End Using a Multi-Phase Gated-Inverted-Ring Oscillator Quantizer. *IEEE Transactions on Biomedical Circuits and Systems 15*, 6 (2021), 1283–1294.
111. Puurtinen, M. M., Komulainen, S. M., Kauppinen, P. K., Malmivuo, J. A. V., and Hyttinen, J. A. K. Measurement of noise and impedance of dry and wet textile electrodes, and textile electrodes with hydrogel. In *2006 International Conference of the IEEE Engineering in Medicine and Biology Society* (2006), pp. 6012–6015.
112. Pérez-Prieto, N., Fiorelli, R., Valtierra, J. L., Pérez-García, P., Delgado-Restituto, M., and Rodríguez-Vázquez, A 32 Input Multiplexed Channel Analog Front-End with Spatial Delta Encoding Technique and Differential Artifacts Compression. In *2019 IEEE Biomedical Circuits and Systems Conference (BioCAS)* (2019), pp. 1–4.
113. Ramkaj, A. T., Peña Ramos, J. C., Pelgrom, M. J. M., Steyaert, M. S. J., Verhelst, M., and Tavernier, F. A 5-GS/s 158.6-mW 9.4-ENOB Passive-Sampling Time-Interleaved Three-Stage Pipelined-SAR ADC With Analog-Digital Corrections in 28-nm CMOS. *IEEE Journal of Solid-State Circuits 55*, 6 (2020), 1553–1564.
114. Ramkaj, A. T., Strackx, M., Steyaert, M. S. J., and Tavernier, F. A 1.25-GS/s 7-b SAR ADC With 36.4-dB SNDR at 5 GHz Using Switch-Bootstrapping, USPC DAC and Triple-Tail Comparator in 28-nm CMOS. *IEEE Journal of Solid-State Circuits 53*, 7 (2018), 1889–1901.
115. Ranjandish, R., and Schmid, A. A Sub- μW /Channel, 16-Channel Seizure Detection and Signal Acquisition SoC Based on Multichannel Compressive Sensing. *IEEE Transactions on Circuits and Systems II: Express Briefs 65*, 10 (2018), 1400–1404.
116. Razavi, B. The Bootstrapped Switch [A Circuit for All Seasons]. *IEEE Solid-State Circuits Magazine 7*, 3 (2015), 12–15.

117. Razavi, B. The StrongARM Latch [A Circuit for All Seasons]. *IEEE Solid-State Circuits Magazine 7*, 2 (2015), 12–17.
118. Razavi, B. Jitter-Power Trade-Offs in PLLs. *IEEE Transactions on Circuits and Systems I: Regular Papers 68*, 4 (2021), 1381–1387.
119. Rovere, G., Fateh, S., and Benini, L. A 2.2-μW Cognitive Always-On Wake-Up Circuit for Event-Driven Duty-Cycling of IoT Sensor Nodes. *IEEE Journal on Emerging and Selected Topics in Circuits and Systems 8*, 3 (2018), 543–554.
120. Saberi, M., Lotfi, R., Mafinezhad, K., and Serdijn, W. A. Analysis of Power Consumption and Linearity in Capacitive Digital-to-Analog Converters Used in Successive Approximation ADCs. *IEEE Transactions on Circuits and Systems I: Regular Papers 58*, 8 (2011), 1736–1748.
121. Safa, A., Corradi, F., Keuninckx, L., Ocket, I., Bourdoux, A., Catthoor, F., and Gielen, G. G. E. Improving the Accuracy of Spiking Neural Networks for Radar Gesture Recognition Through Preprocessing. *IEEE Transactions on Neural Networks and Learning Systems 34*, 6 (2023), 2869–2881.
122. Safa, A., Sahli, H., Bourdoux, A., Ocket, I., Catthoor, F., and Gielen, G. G. E. Learning Event-based Spatio-Temporal Feature Descriptors via Local Synaptic Plasticity: A Biologically-realistic Perspective of Computer Vision. *CoRR abs/2111.00791* (2021).
123. Safa, A., Van Assche, J., Frenkel, C., Bourdoux, A., Catthoor, F., and Gielen, G. G. Exploring Information-Theoretic Criteria to Accelerate the Tuning of Neuromorphic Level-Crossing ADCs. In *Neuro-Inspired Computational Elements Conference (NICE 2023)* (2023), pp. 63–70.
124. Salam, M. T., Kassiri, H., Soltani, N., He, H., Perez Velazquez, J. L., and Genov, R. Tradeoffs between wireless communication and computation in closed-loop implantable devices. In *2016 IEEE International Symposium on Circuits and Systems (ISCAS)* (2016), pp. 1838–1841.
125. Sayiner, N., Sorensen, H., and Viswanathan, T. A level-crossing sampling scheme for A/D conversion. *IEEE Transactions on Circuits and Systems II: Analog and Digital Signal Processing 43*, 4 (1996), 335–339.
126. Schell, B., and Tsividis, Y. A Continuous-Time ADC/DSP/DAC System With No Clock and With Activity-Dependent Power Dissipation. *IEEE Journal of Solid-State Circuits 43*, 11 (2008), 2472–2481.
127. Schinkel, D., Mensink, E., Klumperink, E., van Tuijl, E., and Nauta, B. A Double-Tail Latch-Type Voltage Sense Amplifier with 18 ps Setup+Hold Time. In *2007 IEEE International Solid-State Circuits Conference. Digest of Technical Papers* (2007), pp. 314–605.
128. Schreier, R., and Temes, G. C. *Understanding Delta-Sigma Data Converters*. Wiley, 2005.
129. Scolari, N., Pengg, F. X., Manetakis, K., Salazar, C. A., Vouilloz, A., Serna, E. P., Dissanayake, A., Persechini, P., Kopta, V., Roux, E. L., Chicco, F., Cillo, S., Gerber, N., Barbelenet, C., Epifano, F., Fabbro, P. A. D., and Raemy, N. 23.2 A 1 mm^2 Software-Defined Dual-Mode Bluetooth Transceiver with 10 dBm Maximum TX Power and −98.2 dBm Sensitivity 2.96 mW RX Power at 1 Mb/s. In *2024 IEEE International Solid-State Circuits Conference (ISSCC)* (2024), vol. 67, pp. 402–404.
130. Serteyn, A., Vullings, R., Meftah, M., and Bergmans, J. W. M. Motion Artifacts in Capacitive ECG Measurements: Reducing the Combined Effect of DC Voltages and Capacitance Changes Using an Injection Signal. *IEEE Transactions on Biomedical Engineering 62*, 1 (2015), 264–273.
131. Shaeri, M. A., Shin, U., Yadav, A., Caramellino, R., Rainer, G., and Shoaran, M. 33.3 MiBMI: A 192/512-Channel 2.46 mm^2 Miniaturized Brain-Machine Interface Chipset Enabling 31-Class Brain-to-Text Conversion Through Distinctive Neural Codes. In *2024 IEEE International Solid-State Circuits Conference (ISSCC)* (2024), vol. 67, pp. 546–548.

132. Sharifshazileh, M., Burelo, K., Sarnthein, J., and Indiveri, G. An electronic neuromorphic system for real-time detection of high frequency oscillations (HFO) in intracranial EEG. *Nature Communications 12* (05 2021), 3095.
133. Sharifshazileh, M., and Indiveri, G. An Adaptive Event-based Data Converter for Always-on Biomedical Applications at the Edge. In *2023 IEEE International Symposium on Circuits and Systems (ISCAS)* (2023), pp. 1–5.
134. Shi, W., Cao, J., Zhang, Q., Li, Y., and Xu, L. Edge Computing: Vision and Challenges. *IEEE Internet of Things Journal 3*, 5 (2016), 637–646.
135. Shibata, H. Continuous-Time Pipelined ADC: A Breed of Continuous-Time ADCs for Wideband Data Conversion. *IEEE Open Journal of the Solid-State Circuits Society 3* (2023), 162–173.
136. Shin, U., Ding, C., Zhu, B., Vyza, Y., Trouillet, A., Revol, E. C. M., Lacour, S. P., and Shoaran, M. NeuralTree: A 256-Channel 0.227-μJ/Class Versatile Neural Activity Classification and Closed-Loop Neuromodulation SoC. *IEEE Journal of Solid-State Circuits 57*, 11 (2022), 3243–3257.
137. Shoaib, M., Jha, N. K., and Verma, N. A compressed-domain processor for seizure detection to simultaneously reduce computation and communication energy. In *Proceedings of the IEEE 2012 Custom Integrated Circuits Conference* (2012), pp. 1–4.
138. Shoaran, M., Kamal, M. H., Pollo, C., Vandergheynst, P., and Schmid, A. Compact Low-Power Cortical Recording Architecture for Compressive Multichannel Data Acquisition. *IEEE Transactions on Biomedical Circuits and Systems 8*, 6 (2014), 857–870.
139. Shoeb, A. H. *Application of Machine Learning to Epileptic Seizure Onset Detection and Treatment (Ph.D. Dissertation)*. Harvard-MIT Division of Health Sciences and Technology, Cambridge, Massachusetts, United States of America, 2009.
140. Shrestha, S. B., and Orchard, G. SLAYER: spike layer error reassignment in time. In *NIPS'18: Proceedings of the 32nd International Conference on Neural Information Processing Systems* (2018), pp. 1419–1428.
141. Song, S., Konijnenburg, M., van Wegberg, R., Xu, J., Ha, H., Sijbers, W., Stanzione, S., Biswas, D., Breeschoten, A., Vis, P., van Liempd, C., van Hoof, C., and van Helleputte, N. A 769 μW Battery-Powered Single-Chip SoC With BLE for Multi-Modal Vital Sign Monitoring Health Patches. *IEEE Transactions on Biomedical Circuits and Systems 13*, 6 (2019), 1506–1517.
142. Statista Market Insights. Internet of Things: market data & analysis , 2023. https://www.statista.com/study/109197/internet-of-things-market-outlook-report/ (Accessed: 27-11-2023).
143. Steyaert, M., and Sansen, W. A micropower low-noise monolithic instrumentation amplifier for medical purposes. *IEEE Journal of Solid-State Circuits 22*, 6 (1987), 1163–1168.
144. Tang, W., Osman, A., Kim, D., Goldstein, B., Huang, C., Martini, B., Pieribone, V. A., and Culurciello, E. Continuous Time Level Crossing Sampling ADC for Bio-Potential Recording Systems. *IEEE Transactions on Circuits and Systems I: Regular Papers 60*, 6 (2013), 1407–1418.
145. Tang, X., Renteria-Pinon, M., and Tang, W. Dynamic Predictive Sampling Analog to Digital Converter for Sparse Signal Sensing. *IEEE Transactions on Circuits and Systems II: Express Briefs 70*, 7 (2023), 2360–2364.
146. Tian, F., Wang, X., Chen, J., Zheng, J., Wu, H., Liu, X., Tu, F., Yang, J., Sawan, M., Tsui, C.-Y., and Cheng, K.-T. T. BIOS: A 40 nm Bionic Sensor-defined 0.47 pJ/SOP, 268.7 TSOPs/W Configurable Spiking Neuron-in-Memory Processor for Wearable Healthcare. In *ESSCIRC 2023- IEEE 49th European Solid State Circuits Conference (ESSCIRC)* (2023), pp. 225–228.
147. Timmermans, M., Fattori, M., Harpe, P., Liu, Y.-H., and Cantatore, E. A 3–320 fJ/conv.step Continuous Time Level Crossing ADC with Dynamic Self-Biasing Comparators Achieving 61.4 dB-SNDR. In *2023 IEEE Symposium on VLSI Technology and Circuits (VLSI Technology and Circuits)* (2023), pp. 1–2.

148. Timmermans, M., van Oosterhout, K., Fattori, M., Harpe, P., Liu, Y.-H., and Cantatore, E. A 1.8–65 fj/conv.-step 64-db SNDR continuous-time level crossing ADC exploiting dynamic self-biasing comparators. *IEEE Journal of Solid-State Circuits* (2024), 1–10.
149. Trakimas, M., and Sonkusale, S. R. An Adaptive Resolution Asynchronous ADC Architecture for Data Compression in Energy Constrained Sensing Applications. *IEEE Transactions on Circuits and Systems I: Regular Papers 58*, 5 (2011), 921–934.
150. Tran, C., Kawaguchi, H., and Sakurai, T. Low-power high-speed level shifter design for block-level dynamic voltage scaling environment. In *2005 International Conference on Integrated Circuit Design and Technology, 2005. ICICDT 2005.* (2005), pp. 229–232.
151. Tsai, C.-W., Jiang, R., Zhang, L., Zhang, M., and Yoo, J. Seizure-Cluster-Inception CNN (SciCNN): A Patient-Independent Epilepsy Tracking SoC With 0-Shot-Retraining. *IEEE Transactions on Biomedical Circuits and Systems 17*, 6 (2023), 1202–1213.
152. Tsividis, Y. Event-Driven Data Acquisition and Digital Signal Processing–A Tutorial. *IEEE Transactions on Circuits and Systems II: Express Briefs 57*, 8 (2010), 577–581.
153. Uran, A., Ture, K., Aprile, C., Trouillet, A., Fallegger, F., Revol, E. C. M., Emami, A., Lacour, S. P., Dehollain, C., Leblebici, Y., and Cevher, V. A 16-Channel Neural Recording System-on-Chip With CHT Feature Extraction Processor in 65-nm CMOS. *IEEE Journal of Solid-State Circuits 57*, 9 (2022), 2752–2763.
154. Van Assche, J., Carlino, M. F., Alea, M. D., Massaioli, S., and Gielen, G. From sensor to inference: end-to-end chip design for wearable and implantable biomedical applications. In *2023 IEEE Biomedical Circuits and Systems Conference (BioCAS)* (2023), pp. 1–5.
155. Van Assche, J., and Gielen, G. Power Efficiency Comparison of Event-Driven and Fixed-Rate Signal Conversion and Compression for Biomedical Applications. *IEEE Transactions on Biomedical Circuits and Systems 14*, 4 (2020), 746–756.
156. Van Assche, J., and Gielen, G. A 10.4-ENOB 0.92–5.38 μW Event-Driven Level-Crossing ADC with Adaptive Clocking for Time-Sparse Edge Applications. In *ESSCIRC 2022- IEEE 48th European Solid State Circuits Conference (ESSCIRC)* (2022), pp. 261–264.
157. Van Assche, J., Helsen, R., and Gielen, G. EffiCSense: an Architectural Pathfinding Framework for Energy-Constrained Sensor Applications. In *2022 Design, Automation & Test in Europe Conference & Exhibition (DATE)* (2022), pp. 136–141.
158. Van Helleputte, N., Kim, S., Kim, H., Kim, J. P., Van Hoof, C., and Yazicioglu, R. F. A 160 μA Biopotential Acquisition IC With Fully Integrated IA and Motion Artifact Suppression. *IEEE Transactions on Biomedical Circuits and Systems 6*, 6 (2012), 552–561.
159. Van Helleputte, N., Konijnenburg, M., Pettine, J., Jee, D.-W., Kim, H., Morgado, A., Van Wegberg, R., Torfs, T., Mohan, R., Breeschoten, A., de Groot, H., Van Hoof, C., and Yazicioglu, R. F. A 345 μW Multi-Sensor Biomedical SoC With Bio-Impedance, 3-Channel ECG, Motion Artifact Reduction, and Integrated DSP. *IEEE Journal of Solid-State Circuits 50*, 1 (2015), 230–244.
160. Vanhoof, B., and Dehaene, W. SRAM with stability monitoring and body bias tuning for biomedical applications. *IEEE Solid-State Circuits Letters 5* (2022), 29–32.
161. Verhelst, M., and Bahai, A. Where Analog Meets Digital: Analog-to-Information Conversion and Beyond. *IEEE Solid-State Circuits Magazine 7*, 3 (2015), 67–80.
162. Vezyrtzis, C., and Tsividis, Y. Processing of signals using level-crossing sampling. In *2009 IEEE International Symposium on Circuits and Systems* (2009), pp. 2293–2296.
163. Wang, H., Nguyen, V., Schembari, F., and Staszewski, R. B. An Adaptive-Resolution Quasi-Level-Crossing Delta Modulator With VCO-Based Residue Quantizer. *IEEE Transactions on Circuits and Systems II: Express Briefs 67*, 12 (2020), 2828–2832.
164. Wang, H., Schembari, F., and Staszewski, R. B. An Event-Driven Quasi-Level-Crossing Delta Modulator Based on Residue Quantization. *IEEE Journal of Solid-State Circuits 55*, 2 (2020), 298–311.

165. WANG, T., WANG, D., HURST, P. J., LEVY, B. C., AND LEWIS, S. H. A Level-Crossing Analog-to-Digital Converter With Triangular Dither. *IEEE Transactions on Circuits and Systems I: Regular Papers 56*, 9 (2009), 2089–2099.
166. WANG, Y.-Z., WANG, Y.-P., WU, Y.-C., AND YANG, C.-H. A 12.6 mW, 573–2901 kS/s Reconfigurable Processor for Reconstruction of Compressively Sensed Physiological Signals. *IEEE Journal of Solid-State Circuits 54*, 10 (2019), 2907–2916.
167. WANG, Z., LIU, Y., ZHOU, P., TAN, Z., FAN, H., ZHANG, Y., SHEN, L., RU, J., WANG, Y., YE, L., AND HUANG, R. A 148-nW Reconfigurable Event-Driven Intelligent Wake-Up System for AIoT Nodes Using an Asynchronous Pulse-Based Feature Extractor and a Convolutional Neural Network. *IEEE Journal of Solid-State Circuits 56*, 11 (2021), 3274–3288.
168. WANG, Z., ZHANG, H., ZHANG, Y., SHEN, L., RU, J., FAN, H., TAN, Z., WANG, Y., YE, L., AND HUANG, R. A Software-Defined Always-On System With 57–75-nW Wake-Up Function Using Asynchronous Clock-Free Pipelined Event-Driven Architecture and Time-Shielding Level-Crossing ADC. *IEEE Journal of Solid-State Circuits 56*, 9 (2021), 2804–2816.
169. WEI, D. *Sampling Based on Local Bandwidth (M.Sc. Dissertation)*. MASSACHUSETTS INSTITUTE OF TECHNOLOGY, Cambridge, Massachusetts, United States of America, 2007.
170. WELTIN-WU, C. *Design and Optimization of Low-power Level-crossing ADCs (Ph.D. Dissertation)*. Columbia University, New York, United States of America, 2012.
171. WELTIN-WU, C., AND TSIVIDIS, Y. An Event-driven Clockless Level-Crossing ADC With Signal-Dependent Adaptive Resolution. *IEEE Journal of Solid-State Circuits 48*, 9 (2013), 2180–2190.
172. WERBOS, P. Backpropagation through time: what it does and how to do it. *Proceedings of the IEEE 78*, 10 (1990), 1550–1560.
173. WU, H., CHEN, J., LIU, X., ZOU, W., YANG, J., AND SAWAN, M. An Energy-Efficient Small-Area Configurable Analog Front-End Interface for Diverse Biosignals Recording. *IEEE Transactions on Biomedical Circuits and Systems 17*, 4 (2023), 818–830.
174. WU, T.-F., AND CHEN, M. S.-W. A Subranging-Based Nonuniform Sampling ADC With Sampling Event Filtering. *IEEE Solid-State Circuits Letters 1*, 4 (2018), 78–81.
175. WU, T.-F., AND CHEN, M. S.-W. A Noise-Shaped VCO-Based Nonuniform Sampling ADC With Phase-Domain Level Crossing. *IEEE Journal of Solid-State Circuits 54*, 3 (2019), 623–635.
176. WU, T.-F., AND CHEN, M. S.-W. 16.7 A 40 MHz-BW 76.2 dB/78.0 dB SNDR/DR Noise-Shaping Nonuniform Sampling ADC with Single Phase-Domain Level Crossing and Embedded Nonuniform Digital Signal Processor in 28 nm CMOS. In *2020 IEEE International Solid-State Circuits Conference - (ISSCC)* (2020), pp. 262–264.
177. WU, T.-F., DEY, S., AND CHEN, M. S.-W. A Nonuniform Sampling ADC Architecture With Reconfigurable Digital Anti-Aliasing Filter. *IEEE Transactions on Circuits and Systems I: Regular Papers 63*, 10 (2016), 1639–1651.
178. WU, T.-F., HO, C.-R., AND CHEN, M. S.-W. A Flash-Based Non-Uniform Sampling ADC With Hybrid Quantization Enabling Digital Anti-Aliasing Filter. *IEEE Journal of Solid-State Circuits 52*, 9 (2017), 2335–2349.
179. XU, J., MITRA, S., MATSUMOTO, A., PATKI, S., VAN HOOF, C., MAKINWA, K. A. A., AND YAZICIOGLU, R. F. A Wearable 8-Channel Active-Electrode EEG/ETI Acquisition System for Body Area Networks. *IEEE Journal of Solid-State Circuits 49*, 9 (2014), 2005–2016.
180. XU, J., MITRA, S., VAN HOOF, C., YAZICIOGLU, R. F., AND MAKINWA, K. A. A. Active Electrodes for Wearable EEG Acquisition: Review and Electronics Design Methodology. *IEEE Reviews in Biomedical Engineering 10* (2017), 187–198.
181. XU, X., CAI, Q., ZHAO, Y., WANG, G., ZHAO, L., AND LIAN, Y. A 2.66 μW Clinician-Like Cardiac Arrhythmia Watchdog Based on P-QRS-T for Wearable Applications. *IEEE Transactions on Biomedical Circuits and Systems 16*, 5 (2022), 793–806.
182. XU, X., SUO, Y., ZHOU, P., HAN, X., CAI, Q., WANG, G., LIAN, Y., AND ZHAO, Y. A 2.67G 454 nVrms 14.9 W Dry-Electrode Enabled ECG-on-Chip with Arrhythmia Detection. In *2023 IEEE Custom Integrated Circuits Conference (CICC)* (2023), pp. 1–2.

183. Yan, P., Akhoundi, A., Shah, N. P., Tandon, P., Muratore, D. G., Chichilnisky, E. J., and Murmann, B. Data Compression Versus Signal Fidelity Tradeoff in Wired-OR Analog-to-Digital Compressive Arrays for Neural Recording. *IEEE Transactions on Biomedical Circuits and Systems 17*, 4 (2023), 754–767.
184. Yang, Y., Zhou, J., Liu, X., Cheong, J. H., and Goh, W. L. A 151-nW Adaptive Delta-Sampling ADC for Ultra-Low Power Sensing Applications. *IEEE Transactions on Circuits and Systems II: Express Briefs 63*, 7 (2016), 638–642.
185. Yazicioglu, R. F., Kim, S., Torfs, T., Kim, H., and Van Hoof, C. A 30 μ W Analog Signal Processor ASIC for Portable Biopotential Signal Monitoring. *IEEE Journal of Solid-State Circuits 46*, 1 (2011), 209–223.
186. Yik, J., den Berghe, K. V., den Blanken, D., Bouhadjar, Y., Fabre, M., Hueber, P., Kleyko, D., Pacik-Nelson, N., Sun, P.-S. V., Tang, G., Wang, S., Zhou, B., Ahmed, S. H., Joseph, G. V., Leto, B., Micheli, A., Mishra, A. K., Lenz, G., Sun, T., Ahmed, Z., Akl, M., Anderson, B., Andreou, A. G., Bartolozzi, C., Basu, A., Bogdan, P., Bohte, S., Buckley, S., Cauwenberghs, G., Chicca, E., Corradi, F., de Croon, G., Danielescu, A., Daram, A., Davies, M., Demirag, Y., Eshraghian, J., Fischer, T., Forest, J., Fra, V., Furber, S., Furlong, P. M., Gilpin, W., Gilra, A., Gonzalez, H. A., Indiveri, G., Joshi, S., Karia, V., Khacef, L., Knight, J. C., Kriener, L., Kubendran, R., Kudithipudi, D., Liu, Y.-H., Liu, S.-C., Ma, H., Manohar, R., Margarit-Taulé, J. M., Mayr, C., Michmizos, K., Muir, D., Neftci, E., Nowotny, T., Ottati, F., Ozcelikkale, A., Panda, P., Park, J., Payvand, M., Pehle, C., Petrovici, M. A., Pierro, A., Posch, C., Renner, A., Sandamirskaya, Y., Schaefer, C. J., van Schaik, A., Schemmel, J., Schmidgall, S., Schuman, C., sun Seo, J., Sheik, S., Shrestha, S. B., Sifalakis, M., Sironi, A., Stewart, M., Stewart, K., Stewart, T. C., Stratmann, P., Timcheck, J., Tömen, N., Urgese, G., Verhelst, M., Vineyard, C. M., Vogginger, B., Yousefzadeh, A., Zohora, F. T., Frenkel, C., and Reddi, V. J. NeuroBench: A framework for benchmarking neuromorphic computing algorithms and systems, 2024.
187. Yoo, J. Body Area Network: Connecting and powering things together around the human body. *IEEE Solid-State Circuits Magazine 15*, 2 (2023), 49–58.
188. Zhang, D. *Ultra-Low-Power Analog-to-Digital Converters for Medical Applications (Ph.D. Dissertation)*. Linköping University, Linköping, Sweden, 2014.
189. Zhang, D., Bhide, A., and Alvandpour, A. A 53-nW 9.1-ENOB 1-kS/s SAR ADC in 0.13-μm CMOS for Medical Implant Devices. *IEEE Journal of Solid-State Circuits 47*, 7 (2012), 1585–1593.
190. Zhang, X., and Lian, Y. A 300-mV 220-nW Event-Driven ADC With Real-Time QRS Detection for Wearable ECG Sensors. *IEEE Transactions on Biomedical Circuits and Systems 8*, 6 (2014), 834–843.
191. Zhang, Y., Xu, L., van der Ven, M., Ouzounov, S., Meftah, M., Mischi, M., Cantatore, E., and Harpe, P. A Front-End Amplifier With Fast Reset Scheme for Capacitive Heart Rate Monitoring. *IEEE Sensors Journal 23*, 15 (2023), 17025–17033.
192. Zhao, S., Yang, J., Wang, J., Fang, C., Liu, T., Zhang, S., and Sawan, M. A 0.99-to-4.38 uJ/class Event-Driven Hybrid Neural Network Processor for Full-Spectrum Neural Signal Analyses. *IEEE Transactions on Biomedical Circuits and Systems 17*, 3 (2023), 598–609.
193. Zhao, W., Sun, B., Wu, T., and Yang, Z. On-Chip Neural Data Compression Based On Compressed Sensing With Sparse Sensing Matrices. *IEEE Transactions on Biomedical Circuits and Systems 12*, 1 (2018), 242–254.
194. Zhu, B., Shin, U., and Shoaran, M. Closed-Loop Neural Prostheses with On-Chip Intelligence: A Review and a Low-Latency Machine Learning Model for Brain State Detection. *IEEE Transactions on Biomedical Circuits and Systems 15*, 5 (2021), 877–897.

Index

J. Van Assche, G. Gielen, *Event-Driven Circuit Architectures for Scalable and Adaptive Sensor Readout*, Analog Circuits and Signal Processing,
https://doi.org/10.1007/978-3-031-97285-0

The manufacturer's authorised representative in the EU is Springer Nature Customer Service Centre GmbH, Europaplatz 3, 69115 Heidelberg, Germany. If you have any concerns regarding our products, please contact ProductSafety@springernature.com

Printed and bound by CPI Group (UK) Ltd, Croydon, CR0 4YY
01/10/2026
02225864-0004